EMPIRE
OF ANTS

THE HIDDEN WORLDS AND EXTRAORDINARY
LIVES OF EARTH'S TINY CONQUERORS

SUSANNE FOITZIK
and OLAF FRITSCHE

Translated by AYÇA TÜRKOĞLU

THE EXPERIMENT
NEW YORK

The Experiment, LLC
220 East 23rd Street, Suite 600
New York, NY 10010-4658
theexperimentpublishing.com

THE EXPERIMENT and its colophon are registered trademarks of The Experiment, LLC. Many of the designations used by manufacturers and sellers to distinguish their products are claimed as trademarks. Where those designations appear in this book and The Experiment was aware of a trademark claim, the designations have been capitalized.

The Experiment's books are available at special discounts when purchased in bulk for premiums and sales promotions as well as for fund-raising or educational use. For details, contact us at info@theexperimentpublishing.com.

Library of Congress Cataloging-in-Publication Data

Names: Foitzik, Susanne, author. | Fritsche, Olaf, 1967- author. |
    Türkoğlu, Ayça, translator.
Title: Empire of ants : the hidden worlds and extraordinary lives of earth's
    tiny conquerors / Susanne Foitzik with Olaf Fritsche, translated by
    Ayça Türkoğlu.
Other titles: Weltmacht auf sechs Beinen English
Description: New York, NY : The Experiment, 2021. | Originally published in
    Germany as Weltmacht auf sechs Beinen by Rowohlt Verlag GmbH, Hamburg,
    in 2019.
Identifiers: LCCN 2020047939 (print) | LCCN 2020047940 (ebook) | ISBN
    9781615197125 | ISBN 9781615197132 (ebook)
Subjects: LCSH: Ants.
Classification: LCC QL568.F7 F5413 2021 (print) | LCC QL568.F7 (ebook) |
    DDC 595.79/6--dc23
LC record available at https://lccn.loc.gov/2020047939
LC ebook record available at https://lccn.loc.gov/2020047940

ISBN 978-1-61519-712-5
Ebook ISBN 978-1-61519-713-2

Jacket and text design by Jack Dunnington
Cover photograph by Alex Wild
Illustrations by Susanne Foitzik
Author photographs by Peter Pulkowski (Susanne Foitzik)
and courtesy of Olaf Fritsche

Manufactured in China

First printing March 2021
10 9 8 7 6 5 4 3 2 1

CONTENTS

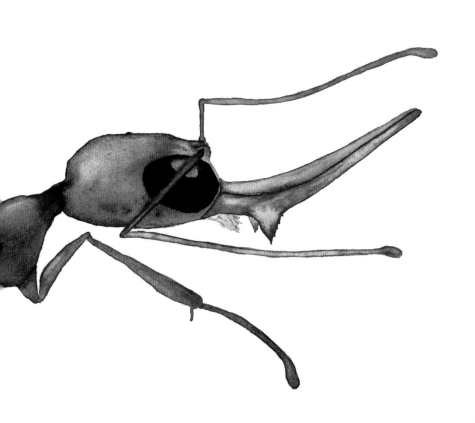

SMALL BUT MIGHTY

The Indian ant species Harpegnathos saltator *uses its large eyes to hunt.*

They often say people look like ants when seen from a great height.

I lean forward a little in my seat and look out of the window. Our flight to Peru has yet to reach cruising altitude. Below us, I can just about make out roads, houses and fields, cows in a pasture, here and there a village, and a city in the far distance. It's not a bad analogy, I think. Everything I can see from up here is something I have encountered in my research on ants: fixed roads, spectacular buildings, agriculture, livestock farming. I let myself sink back into my seat. The parallels don't stop there, if you think about it. Ants live in cities, just like people. In times of peace, they go about their work, sharing the labor fairly. Each has a job, from the wet nurses in the nursery to the architects, construction workers, and housekeepers in the nest, to the hunters and gatherers, who ensure that everyone is fed. But peace does not last forever, even among ants. Neighbors fight over the borders to their territory and wage bitter wars on one another. Invaders overwhelm unsuspecting colonies. The weaker members are carried off and enslaved. Ants look on as whole empires rise and fall.

Just like us.

NOT ALL QUEENS ARE CREATED EQUAL

Of course, the parallels don't necessarily mean that ants and humans are the same, or that ants would do a better job of being human than we do. That kind of thinking is way off the mark. Many similarities have only emerged because we have applied terms of human social structures to the organizational structures of ants. Take, for example, ant queens. They have almost nothing in common with human monarchs—unless you know of any queens who literally gave birth to all of her subjects. And you won't encounter anything like your average female ant worker in a human factory. Yet we use the same descriptors in both situations, because we have no better words for them, and new terms would make things needlessly complicated. Plus, these terms are surprisingly apt when it comes to describing the attributes of ants. But we should bear in mind that these terms don't mean the same thing when applied to human societies.

It's impossible to deny the parallels between people and ants, but these commonalities have emerged in different ways—for ants, they are solely the result of natural selection; for us, they are also the result of civilization, technology, and a system of moral values, but we often find similar problems confronting us: How can a large number of individuals live together in a small space? How do groups compete for resources? How do we stand our ground in an environment full of potential dangers?

Commonalities and differences: Perhaps it is this combination of the two that makes ants so fascinating to us.

STAGGERINGLY SMALL

The most striking difference between ants and people is size: Ants are obviously much smaller than us. The largest specimens can reach the size of a june bug and the smallest are barely the size of the dot on the letter "i." The tiny pharaoh ant (*Monomorium pharaonis*) could crawl about on the head of a carpenter ant (*Camponotus gigas*) without much trouble—a size difference akin to that of a man and a mouse.

In human terms, the "nations" of ants we've observed have been as tiny as San Marino and as massive as China. Colonies of the genus *Temnothorax* comprise just a few dozen insects and fit neatly inside an acorn. By contrast, colonies of leaf-cutter ants can number as many as three million, living in underground nests as big as houses. These are outfitted with extensive rooms, corridors, ventilation systems, humidity control, waste disposal facilities, and climatic chambers for growing fungus. These are cities of a million insect inhabitants, much like New York, but built entirely underground. Would humans be capable of creating and sustaining a similar underground world?

Amid all these comparisons, we must not forget that when we talk about humans, we are only talking about a single species, *Homo sapiens*, whereas the world of ants boasts thousands of different species, some of which indulge in very different lifestyles. There are variations and exceptions to almost every pattern of behavior we encounter in this book, some of which only appear in a few species. But in many ways, (almost) all ants are the same. And it is these commonalities to which they owe their unique place in the world.

THE SECRET TO SUCCESS

The secret to ants' success is largely grounded in their attitude to life: "Ask not what your colony can do for you, ask what you can do for your colony." An ant is helpless without its colony but is always ready to sacrifice itself on the colony's behalf. Even the mighty ant workers of the tropical bullet ant species *Paraponera clavata*, which can grow as large as hornets and have a sting so painful it can drive a person insane, die within a few days of losing their colony. But when ants work as a team, they are practically unstoppable. It's why locals abandon their huts when a colony of tropical army ants arrives to raid their farmsteads or villages. Pity the forgotten pets, left tied up or locked inside by their owners.

Another secret to their success is the sheer number of them. Nobody knows how many ants there are in the world. Some scientists estimate that there could be around ten quadrillion. If that's correct, then there are a million ants for every human on the planet. Assuming that these little insects are, on average, less than half an inch (1 cm) long, if you lined up all the ants in the world, they could form a chain reaching from the sun to Earth and back 344 times. Or ten times further, as far as Pluto, once the most distant planet in the solar system. These fun figures suggest that each ant's miniscule contribution is amplified by their sheer quantity, making them hard to ignore. An individual ant might not eat much, but altogether ants destroy countless tons of other insects every year. An individual ant might shift a few grains of sand, but together they can move mountains. And they have been doing it since they emerged over one hundred million years ago, when dinosaurs still ruled Earth.

WHO REALLY RUNS THE WORLD?

These days, humans consider themselves rulers of the world. Yet the world would be quite happy without us. Imagine what would happen if all the humans on the planet disappeared. Whole books have been devoted to this topic and they have all concluded that no global catastrophe would occur. On the contrary, in time nature would recover from our reckless reign, reclaiming towns and cities, producing new species, and returning to the state of biodiversity it boasted just a few thousand years ago. Of course, humanity is part of nature, but in many ways our behavior is no longer in balance with its natural cycles. We spew more carbon dioxide into the atmosphere than can be removed by all the world's photosynthetic plants, algae, and microorganisms combined, resulting in the catastrophic greenhouse effect. Elsewhere, our ecosystems are unable to cope with the waste we produce, such as radioactive substances and plastics, which is why they are left to accumulate underground and in our seas. You might say that we humans are more a curse than a blessing for life on Earth.

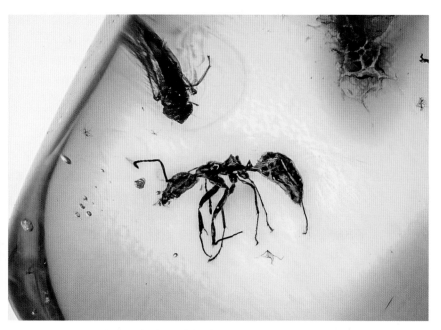

Ants crawled at the feet of the dinosaurs. Sometimes they would become trapped in droplets of resin, which later turned into amber.

Not so with ants. Of course, there are species of ants that create monocultures and significantly change the ecosystems in which they live, but since these species are relatively few in number and are not present across the globe, ants live much more sustainably. Their behaviors have developed more slowly, allowing ecosystems to adjust. Without ants, for example, the species of insects they prey upon would quickly multiply, devouring woods and grasslands. Dead animals would no longer be disposed of and lie rotting where they fell. The soil would be inadequately aerated, nutrient cycles would stall. If all the ants suddenly disappeared, terrestrial ecosystems across the world would be on their knees and it would take a number of years, decades—centuries, even—for them to achieve a new balance. Without ants, the natural world would suffer a long period of instability and would never look the same again. So, who really runs the world?

CREEPY-CRAWLIES GUARANTEED

Trappers, paramedics, slavemakers . . . ants play surprising roles. In this book, my coauthor, Olaf Fritsche, and I will be taking you on a journey through the world of the most fascinating and important creatures on the planet. Our journey will take us to Peru, Malaysia, North America, and the forests of Europe. We'll discover queens who have no say in their own colonies, female bodyguards who hitch a ride on their clients' backs and fend off aggressors, and Amazons who steal other species' offspring just to avoid doing their own chores. We'll explain why some ant queens feed on the blood of their own young and why some workers are prepared to offer up their lives to feed their hungry sisters. We'll also learn why ants don't notice sneaky beetles eating their offspring right under their noses.

You'll be there in the dead of night as I face down a platoon of army ants, when leafcutter ants destroy our lab equipment, and when my colleagues dissect an ant's brain. You'll discover astonishing parallels between ant societies and human ones, and why we are so alike, but also completely different.

Ants: They may be small, but oh, they are mighty!

CHAPTER 1

TAKE ME TO YOUR LEADER!

———

Minor and major workers of the leafcutter ant
Atta cephalotes *differ dramatically in size.*

Bureaucracy is the bane of researchers everywhere. Every expedition to other parts of the globe begins with a mountain of paperwork, which grows taller and taller by the year.

But, luckily, ants can also be found closer to home. If you have ever picked the wrong spot for a picnic, you know only too well that ants will be drawn to sweet treats in no time and have no qualms about inviting themselves to lunch. They can be found in fields, forests, and gardens, and sometimes they even set up shop in our homes. Ants are practically everywhere. They are substantial in number and variety, too. North America is host to nearly 1,000 different ant species. But that's nothing compared to the at least 16,000 different species to be found crawling across the globe. There is also likely to be an unknown number of other ant species, which we have yet to discover because they are rare or live secretive lives. The aim of many researchers is to trace these secretive species, striking out to scour the jungle, steppes, prairies, or deserts, turning over every stone, looking under every leaf, and tapping every branch.

For better or for worse, certain species of ants exert a particular pull on researchers. For example, the secret superstars of the ant world, such as leafcutter ants and army ants, are not to be found in my native Germany. To research these creatures and their fascinating habits in the wild, we (the scientists) must travel to their parts of the world—and that's practically anywhere else on the planet, with the exception of the Antarctic and isolated regions in colder parts of the world, such as Iceland. Anyone wanting to get to know them will often have more forms to fill out than all the legs on all the ants in the world combined. At least, it feels that way.

Much like the objects of our studies, however, those of us who research ants—myrmecologists, as we're called—are pretty persistent. So, every year, in spite of the mounds of paper in our path, we gather together a colorful team of undergrads, PhD students, postdocs, and professors in the most remote corners of the world, facing rattlesnakes, beaded lizards, and leeches, to search for ants and dig them up, before sauntering over to customs where we cheerily present the baffled officials with the countless permits required to ensure smooth departure and arrival for our creepy-crawlies.

Yes, it is at times a headache. Nevertheless, I think it's the best job in the world.

SISTERS ARE DOING IT FOR THEMSELVES

Once we've conquered Paperwork Mountain and touched down near the site, hired the SUV, worked together to push a fallen tree off the road, loaded up the longboat—checking it won't capsize—and found the best route through the current in the falling twilight using the flashlight stored in the bow, it's time to begin the search.

But even ants of the same species differ wildly. An ant society has a strict hierarchy, separating ants into specific castes and sexes: female queens, workers, and males. Queens are at the top of the hierarchy and each colony has one or several queens depending on the species. Her subjects are almost exclusively of female ant workers, who are typically all sisters. By contrast, there are practically no males at all in an ant colony. In the next chapter, I'll explain why this is and why males really don't get much out of life in the ant colony. But first, let's talk about female ants and what it is that makes the small but definite difference between males and females.

When it comes to human beings, Mother Nature, as we know, relies on chromosomes: A fertilized egg with two X chromosomes will develop into a girl and an X and a Y chromosome will produce a boy. Hormones take care of the rest. This is so obvious to us that we tried to believe it was the case for all animals. But nature is much more imaginative than that. Take crocodiles, which let temperature decide: If eggs are incubated at temperatures lower than 86°F (30°C), they will only produce girls; anything above 93°F (34°C) will produce exclusively boys and anything between the two will produce a mix. For clownfish, like Disney's Nemo, it is all a matter of age: All clownfish begin their life as males and, once they mature, they have the ability to change their sex within the space of a week if the dominant female of the group dies. A tropical species of mite, *Brevipalpus phoenicis*, displays an even crazier method of sex determination: due to a bacterium, which causes males to change sex during development, they are all female.

Ants don't let chromosomes decide, either. Instead, everything depends on whether an egg cell has been fertilized. If an egg and a sperm cell merge, they will produce a female ant, if not, the larva will develop into a male. It sounds quite simple but becomes rather more complicated once we look closely at the relationships within a colony (we deal with this in greater detail in chapter 3, Effective Anarchy). We have observed that female ants are more closely related to their sisters than they are to their own sons and daughters, or to their mothers, the queens. Conversely, crocodiles, clownfish, mites, and humans all exhibit the same degree of relatedness between siblings, children, and parents. As I will show you in later chapters, the strange nature of kinship within the ant colony explains a whole range of their no-less-bizarre behavioral abnormalities. Among these is the fact that most female ants do not produce offspring, instead leaving reproduction to the queen.

Yes, that's right: In the ant world, only the queen has the right to a sex life—and even then, it only really happens once. On her "nuptial flight," lasting just a matter of hours, she couples usually with a single male—or more rarely, several—and diligently collects all the sperm she requires for a lifetime of egg-laying. She preserves the semen in a special pouch between her ovaries where it miraculously remains viable for many years (an ant queen can live up to 30 years). Using this stock of semen, the queen can fertilize virtually every egg she lays. And since fertilized eggs produce only female ants, ant colonies are composed almost exclusively of females.

Still, this colony is unfortunately lacking in staff, so the sisters have to manage life on their own. The sisters receive the freshly laid eggs of their mother queen and care for them. They feed the larvae, ensuring that they grow steadily. The sisters keep the offspring clean and keep pathogens away. They renovate the nest and, when necessary, build a new one. They prepare food for everyone and distribute it according to an unfair but strictly organized system. They defend the colony against attackers of all sizes, ready to sacrifice themselves if necessary. In short, the sisters do every and any job required, which is why we typically refer to them not as *sisters* but *workers*.

Ironically, it's the inconspicuous workers who are the most varied and interesting caste in any ant colony. They build magnificent buildings, farm, carry out raids and forays, and wage wars. But these great architects and soldiers begin their careers more modestly than you would imagine: They begin as wet nurses.

A CREEPY KIND OF NURSERY

A typical ant worker spends her whole life trudging back and forth between the nest and the outside world. She begins life as an egg before hatching as a small, white larva that resembles a little worm and is constantly hungry. As soon as she is big enough, the larva transforms into a stiff pupa, which can be enclosed in a cocoon. During this phase, the larva magically changes from her wormlike form into a mature ant.

This process is like the transformation of a caterpillar into a butterfly—but without the wings, and much less colorful. Ant larvae and pupae are what laypeople often refer to as "ant eggs." Yet true eggs are much smaller, with large larvae and pupae being similar in size to their adult relatives. Once the metamorphosis is complete, the ant hatches and gets to work.

While the exoskeleton (cuticle) of the young ant worker is still hardening, she is immediately put to work in the nursery. This means feeding the queen and supporting her with egg-laying. Every egg must be nurtured; newly trained wet nurses must keep the eggs clean and ensure that the temperature and humidity in the nursery remain stable to keep the eggs from freezing or rotting. Both eggs and larvae must be tended to and fed. If you have children, the procedure up to this point might seem somewhat familiar. Diaper-changing is not a problem for ants—at this stage, the larvae have not developed anuses. Anything that goes in one end and isn't used

A family photo of the Carpenter ant, Camponotus discolor. *The large queen (top) jettisons her wings after the nuptial flight while the male (left) dies shortly after mating. The workers (bottom) carry out all work inside the colony with the exception of egg-laying.*

simply collects at the end of the midgut and is known as a meconium. This prevents larvae from soiling one another, keeping the brood chamber clean. Larvae release this fecal matter shortly before pupating, which is why each cocoon bears a natural dark spot.

Now, back to the larvae themselves. Like all young children, baby ants are always hungry. However, these larvae are not the only ones doing the eating—in some ant species, *they* are on the menu. The queens of the American genus *Temnothorax* love lapping up a secretion produced by their larvae. The *Stigmatomma silvestrii* of the Far East, sometimes known as Dracula ants, are even more gruesome: If the queen is hungry, she selects and bites an older larva on the flank until she draws a transparent blood known as hemolymph. She then sucks on the droplets of blood for several minutes before producing a couple of white droppings, which are promptly and discreetly disposed of by her ladies-in-waiting. A meal of blood keeps the queen going for several hours but, like any vampire, she will feed on blood and nothing else. She will spurn tasty prey, even when it is right under her nose. Surprisingly, the larvae she preys on will make a full recovery, bearing just a few scars from where they have served as food. The myrmecologist Keiichi Masuko has found these scars on insects examined in the lab and larvae in nature, which suggests that vampirism is part of a repertoire of normal behaviors for this species and not a phenomenon unique to laboratories.

Still, at least the larvae of vampire ants survive their turns as the Royal Snack. The workers' eggs are not quite so lucky. The young ant workers of most species do possess functional ovaries. However, as long as there is a fertile queen in the nest, they will not reproduce. But they do sometimes lay eggs, none of which develop into little ants, instead becoming ant food. These trophic eggs are a special delicacy for the queen, who depends on the nutrients they provide to produce her own eggs.

In the ant world, "baby food" is unusually meant literally. It may sound horrifying from a human perspective, but it's actually an ingenious survival strategy. Without freezers and twenty-four-hour supermarkets, it is immensely difficult for ants to preserve perishable foods until they are needed. The solution to the problem is live "canned goods," such as eggs, larvae, and pupae. They are sacrificed so that the colony can thrive and grow. If you think about it, these young martyrs benefit indirectly from interfamilial cannibalism: If the colony lives on, so too will their genes.

MISSION: SUICIDE

The time ant workers spend on duty in the nursery can range from a few weeks to a few months, depending on the species. Once this time has elapsed, the next generation is ready to take over. Our female ant worker is not so young anymore and applies herself with renewed vigor to taking care of the household and repairing the nest. She prowls the ant nest and tends to all the jobs she comes across: She maintains the cleanliness of the nest, destroying bacteria and fungi, clears away stones and rubbish, creates new nesting chambers, digs tunnels, keeps ventilation channels clear . . . and sometimes she simply lazes around. Young workers employed inside the hill are guaranteed to take a break and stand around, just for the heck of it. Perhaps it's because there's nothing else to be done at that moment, because as soon as there is, an ant worker will diligently get to work, just as you'd expect. Perhaps there has been such a wealth of food available that the queen has laid vast amounts of eggs and the newly hatched workers simply have little to do but hang around and wait for a job. Whatever the reason for this unexpected laziness, we should probably indulge them. After all, the next step up the career ladder is a veritable suicide mission.

Ant workers do not see daylight until the latter stage of their lives. While out on field duty, they display the behavior that we all associate with ants: scurrying tirelessly along, lugging dead insects with them, crawling around in picnic baskets, and biting us weakly on the calves when we stray too close to their anthills.

While we are angrily rubbing our sore, bitten legs, we forget that life outside the nest is extremely dangerous for ants. Predatory insects such as beetles, wasps, and ants from other colonies will happily pounce on workers, as will spiders, lizards, and many birds. On top of that, there's the risk of desiccation in hot weather, or losing your way on your return to the nest after an excursion that goes on a little too long. That's why desert ants record an average survival time of just two weeks outside the nest. Not long at all, when you consider that some workers can live up to three years of age under laboratory conditions, with the black ant *Formica fusca* living up to eight years. Perhaps this is why ants are always in such a hurry to keep moving when they're out and about.

The risks may be big, but somebody has to provide food for the colony. Young ants have their working lives ahead of them, which is why they are not the natural first choice for exposure to higher risks. This task falls to older workers, giving rise to a common saying among researchers: "Humans send their young men to war, ants send their old ladies." These "old ladies" are hardly old at all; they are still powerful, able-bodied insects and represent a kind of elite troop within the colony. Hormones ensure that the ants leave their home behind and move out into the wild: Hormone production begins when the ant is between a few weeks and a few months old, discouraging them from seeking to care for offspring and tend to the home and instead awakening their sense of adventure. They set out, depending on the species, to milk lice, cut leaves, find dead animals, or hunt prey. They can be found crawling along the ground, on branches and leaves, and this is where we are at last able to spot them.

TRESPASSING IN THE NAME OF SCIENCE

Of course, as ant workers hurry busily back and forth between their food source and the nest, they'd never suspect that the large shadow looming over them was a group of mosquito-bitten researchers almost shaking with excitement at finally having stumbled across them. It's difficult enough keeping track of these little foragers during the day as they make their way secretively through the leaves and undergrowth. However, at night, it's a whole other ball game. It's as if the nests of nocturnal ants are hidden beneath some kind of invisibility cloak: Even with years of experience, they are impossible to find even in bright sunshine. So when we're on the hunt for army ants in Malaysia or leafcutter ants in Peru, we sync our circadian rhythms to those of the little critters: We sleep until midday, if the heat has kept us up, then we write up our report on the previous night's hunt. Then it is time for a quick, strong cup of tea to help us stay awake and after sunset we don bright headlamps and follow the narrow trail into the rainforest.

Sometimes we hear army ants before we see them. A quiet rustling, the sound of hundreds of feet pattering on foliage. Then we notice the ground beneath us moving. When army ants migrate, the entire colony packs its bags (their eggs, larvae, and pupae), forms a broad column and treks to a new nesting site. They spend several hours marching past without paying us the slightest attention. At least, that's what we think at first because it's the rainy season and we're wearing rubber boots. It takes a few minutes before sentry ants are trespassing on our most delicate body parts. At this point, we conclude it's high time we stepped aside to observe the spectacle from a safer distance.

Encounters with leafcutter ants are less likely to end in physical pain. Columns of leafcutters often reach over 260 feet (80 m) long, leading along well-constructed trails through the jungle and straight up a tree. Special cleaning troops ensure that even the smallest twig is cleared from the ants' path, guaranteeing no obstacles or traffic jams. I wish human highway maintenance was as efficient. The ants are not weighed down on their way out, but they return carrying pieces of green leaf like flags in their jaw pincers, or mandibles. Sometimes they also harvest flower petals, creating dancing trails of pink on the forest floor. We will look more closely at the purpose of this sea of flags later, but first we're going to follow the stream of ants to the entrance to the nest. Our goal—for the sake of science, of course—is to gather whole colonies and take them home to our laboratories.

I know it's cruel, but believe me, sometimes there is just no other way. Researchers would love to conduct their investigations out in nature, just as other zoologists do with lions, geese, humpback whales, or sand lizards. Where possible, we follow the ants through jungle, forest, savanna, and desert to observe their behavior in their natural environment. We gladly accept long journeys, hungry mosquitoes, and improvised sanitary facilities, but it's not enough to answer all the important questions. This is partly because ant nests are not see-through, so we cannot see what goes on inside them. Without observing colonies in the lab, we would hardly know anything about how young ants are reared and how they develop from eggs into mature insects.

Humans simply lack the right kind of senses for some studies. In chapter 4, Communicative Sensuality, I will show you how ants communicate using chemical signals, which our noses are unable to pick up. We can prove the existence of these scents using modern methods of analysis, but you try dragging a mass spectrometer through the jungle! The experiments we have performed in our laboratory would not be possible in nature. Unless you thought someone in the forest would eagerly respond to our request to "gather up all the food plants and prey insects and move them just to the left of the nest, please?" Working in the laboratory allows us to create mazes, build obstacle courses, and lay down false scent trails to our heart's content.

If we want to understand, protect, or—in specific cases—control ants, we have no choice but to kidnap a few on occasion. The ants make their views on this quite clear—with plenty of stings and bites. A researcher's life is not always easy.

A PAINFUL MATTER

When it comes to our kidnapping campaigns, it's all or nothing for the ants. And that's how they fight. Strictly speaking, ants are always fighting for all or nothing, even when it costs them their lives, which doesn't make our plans any easier. Their arsenals include powerful jaws, corrosive acids, and sometimes even venomous stings. Yes, many species of ant have stings. The large wood ants of the genus *Formica* or the smaller black turfgrass ant *Lasius neoniger* from North America do not sport stings, though they can spray formic acid, but the invasive red fire ant *Solenopsis invicta* and the invasive European fire ant *Myrmica rubra* use their stings to make their displeasure loud and clear.

You will certainly have discovered for yourself, at some point or other, that ants can be quite unpleasant, so you will no doubt be happy to learn precisely *how* painful your experience was in scientific terms.

US researcher Justin Schmidt specializes in quantitative definition of the sting pain of different insects and, after more than 1,000 painful encounters with over 150 different varieties of stings, has developed a scale, named the Schmidt Sting Pain Index, in his honor. The scale runs from 1.0 to 4.0+ ("+" meaning that the pain exceeds the scale). At the bottom of the scale are stings that are merely unpleasant and forgotten about within five minutes. Schmidt describes these as "almost fruity," comparing them to a somewhat overenthusiastic lover

No sting is more painful than that of the South American bullet ant Paraponera clavata.

nibbling on your earlobe, or a weak electric shock after walking over a flokati rug. The American fire ant *Solenopsis xyloni* registers at this level, whereas ants of the Myrmicinae subfamily pack much more of a punch when they're angry. They include both leafcutter ants and red ants (*Myrmica rubra*): They punish their enemies with a Schmidt Sting Pain Index score of 1.8, which is thought to feel like having a paperclip shot at your cheek. But while you're rubbing the affected area, take comfort in the thought that you haven't been busted by the Australian bulldog ant *Myrmecia gulosa*. Its sting is reported to feel like using a drill to free an ingrown toenail. The pain lasts for several days and scores a 3.0 on the pain index.

Still, that's nothing compared to the sting of the South American *Paraponera clavata*, commonly known as the bullet ant or twenty-four-hour ant. Growing up to an inch and a half (4 cm) in length, it is one of the largest ants in the world and its venom causes the worst pain that Schmidt has ever experienced. He describes this as, "pure, intense, brilliant pain; like walking over flaming charcoal with a 3-inch nail embedded in your heel," and assigns it a 4.0+. Other unhappy test dummies compare the pain to being shot. Thankfully, the sting doesn't cause any lasting damage. After a hellish twenty-four hours, the pain gradually abates. But anyone who has experienced it is unlikely to have the slightest inclination to repeat the experiment, even for the purposes of science. Unless they're one of the Sateré-Mawé people. Men of this tribe in the Amazon Basin participate in a traditional initiation test in which they spend a half hour with their hand submerged in a bag of bullet ants. And they do not just do it once; they do it as many as 25 times.

And they say scientists are crazy.

A BIG HEAD AND A FIRM BITE

Ants that can't sting resort to biting and spraying acid. Ants belonging to the soldier subcaste do this exceptionally well. These are workers with a warlike appearance, noticeably larger than normal workers or minors, which is why we call them majors or supermajors. Their heads are significantly larger and sport powerful jaws. The majors of the genus *Pheidole* have even carried this into their common moniker: "big-headed ants."

The decision as to which ants become the more slight minors and which become the brawny majors is made when the ants are still in the nursery: Larvae that receive less food develop into ants fit for day-to-day business; pampered, well-fed larvae are destined for careers in the military. This unequal treatment is not only reflected in their size, the glands controlling feeding secretions and toxins also grow healthy or stunted depending on their future use. This is all the more surprising given that even the most bellicose Amazons begin their careers shortly after hatching as minor workers do, by working in the nursery. In the ant world, every soldier has to be a wet nurse first.

With their large jaws, majors often take on the job of guarding the entrances to the nest: Leafcutter ant majors use their strength to shred coarse plant material; army ant majors secure the flanks of marching columns and catch and kill large, powerful prey; big-headed ant majors grind hard seeds into crumbs; when knitting together their nests, weaver ant majors carry larvae this way and that, using strands of sticky silk to stick them together; and majors of the genus *Colobopsis* act as doorstops, using their big heads to keep the entrances to the nest tightly closed. Majors are not above returning to the nest to tend to the larvae if, for some reason, the minors responsible for them are suddenly unavailable. With their large jaws, they can sometimes be a little clumsy or lack the requisite gentleness when picking up young larvae. We will be looking more closely at these abilities in the chapters to follow, but we can be certain of one thing: a major in peacetime is not just a high-end ant with bigger muscles.

We come face to face with these little defensive forces when we stick our tools in the entrance to a nest to excavate the colony and abduct its queen. I am sure you will understand that, despite the difference in size and our scientific curiosity, we are happy to reach—respectfully, of course—for our spades, trowels, and knives. What will be, will be.

BAGGED UP AND CARRIED AWAY

Understandably, ants aren't thrilled when we dig up their nests, and I must admit that we aren't especially prissy about it.

When excavating particularly small ants with tiny, acorn-size nests, we spend hours using sharpened penknives to cut through acorns, nuts, and twigs, checking to see if they are occupied. We will sometimes accidentally cut into a worker or even the queen. On occasion, someone will slip and cut their finger, which is why half the team can be found sporting colorful Band-Aids by the end of a weeklong expedition.

Sometimes we wait for a tree to be felled and seize the opportunity to collect all the nests the huge old tree has been host to for decades.

When collecting ants living in the crevices of rocks or flat nests on the ground, we use a different method, which works according to the vacuum cleaner principle. We stick two plastic tubes through a plug, which sits on top of a small collecting glass: This contraption is known as an aspirator. To capture the ants, you hold the end of one of the tubes over the insect and suck through the other tube with all your might. A piece of fine netting stops you from accidently swallowing the ant—provided that you haven't forgotten to attach it, which does happen from time to time. Ants might be a thoroughly acceptable dish on the menu in some cultures, but I can assure you in our experience they are not to everyone's tastes.

It is always particularly exhausting when the object of a researcher's desire has hidden itself deep underground. Then only spades, shovels, and muscle power will do. In Arizona, we spent several hours excavating a colony of the slavemaking ant *Polyergus breviceps*. Exhausted and drenched in sweat, we finally found the brood chamber and the lone queen 5 feet (1.5 m) down. All we had to do was lift it out with one more cut of the spade. But either we were too clumsy or too tired, and the spade missed—slicing the queen in two. After some extensive cursing, we decided to take the colony with us anyway; though it could produce no new workers without a queen, we knew it was our own bad luck that had doomed it to destruction.

When digging out large colonies, we shovel them straight into voluminous plastic tubs and buckets, the rims smeared with super-slippery Fluon. This should prevent any attempted escapes, but as a precaution we also seal the containers with fitted lids during transport. Otherwise it's extremely irritating and distracting being stung and bitten on the calves on the ride home.

Small colonies of tiny ants need little more than a plastic bag filled with some leaf litter and a moist paper towel to keep them going during the journey and stop them from drying out. For food we provide biscuits and tuna, which the ants accept enthusiastically as if they have never known any different. Kept this way, they survive unscathed in the refrigerator at our lodgings for two to three weeks. We can accumulate well over 1,000 bagged colonies over the course of one excursion, so we are always certain to make sure that our lodgings are equipped with very large refrigerators. On one occasion in Michigan, I hadn't bargained on the landlord's wanting to use the apartment while we were there. He was quite surprised to come back from the supermarket, open the fridge and find himself confronted with several hundred colonies of ants. But he took it with good humor, moved some bags aside, and made some space for his milk cartons.

STUNG, BITTEN, AND HAPPY

After an exhausting day traveling or an arduous night collecting ants, it is finally time to return to our lodgings. If we're lucky, it will be a permanent field station with all the usual comforts, such as a well-equipped kitchen, electricity, running water, shower, and even internet. Still, sometimes we have to make do with simple huts and an outhouse. Research in the field might sound romantic, but it stinks—sometimes literally.

Mealtimes are an even bigger adventure. We often cook for ourselves and, with a truly international team, it's as enjoyable a culinary experience as you could expect. Rice, fish, and fresh mangoes are quite delicious—though the fish salad for breakfast in Peru took some getting used to.

What you don't ever get used to in the few weeks you spend on a research trip, however, are the unwelcome roommates that wind up in your bed sheets. In Malaysia, I had draped a net artistically over my bed for extra protection against mosquitoes and had succeeded in keeping the whining bloodsuckers at bay. I had not, however, realized that the forest rats regarded such nets as a helpful climbing frame and only noticed in the middle of the night when I felt the creature's delicate little paws running across my face. And the rats weren't alone. In Malaysia, there are leeches that live on land and which you can pick up when, fascinated for instance by an army ant raid, you crouch down in the rainforest undergrowth. Unfortunately, their bites are almost painless, so you don't notice them.

One morning, I was surprised to find blood stains on the T-shirt I was using as a nightdress in the spots where it had turned out leeches had attached to my skin. The next night, when I went to bed, I noticed that the forest rats had nibbled holes in the fabric where the blood had seeped into the material. From then on, I would arrange my mosquito nets with utmost precision every evening, tucking the ends under the mattress and carrying out a thorough check for hungry leeches before going to sleep.

And yet somehow to me there is still nothing better than being out in nature, looking for ants.

CHAPTER 2

THE BIRTH OF A COLONY

―――――

A young leafcutter queen (Acromyrmex versicolor) establishing a new colony with her first eggs and larvae.

f there's anyone for whom the phrase *L'état, c'est moi* ("I am the state") is true, it's the ant queen. She is the past, present, and future of her colony. She establishes the colony, bears all the ant workers, males, and young queens that live within it, and once she dies, the colony will often perish soon afterward. And to achieve all this, all she has to do is take a trip out her front door.

A QUEEN'S GIRLHOOD

Ant queens are generally made, not born. Excluding a few exceptions, genes rarely play a part in this process. The nannies in the nest nursery follow to a T the recipe for creating "young queens" (as we call them). First, the red wood ants, *Formica obscuripes* for instance, take one of the winter eggs, laid at the end of the dormancy period necessitated by the cold temperatures. This process is not possible with eggs laid in the summer, for reasons that remain a mystery to science, and don't seem to bother the ants much either. But the right egg is just the beginning; the wet-nurse ants must also feed the royal larvae a special diet. This must be plentiful and of high quality. If a red wood ant princess receives this cocktail of nutrients within the first three days of her life as a larva, then the die is cast and she is on the path to becoming a hopeful young queen. And she will be queen—along with a couple hundred of her sisters from the noblest caste.

When it comes to the line of succession, ants do not like to put all their eggs in one basket. The risks of something going wrong are many. From the disappointment of no suitable prince being available or a failure to mate, or the lack of an appropriate location for a new nest nearby, to a deadly encounter with a predator such as a woodpecker—which feast on well-fed young queens—there are any number of opportunities for failure and an early demise. According to estimates, only one in 10,000 young queens is successful in founding a new colony.

The young virgin queens have no inkling of this at first. They are far larger than their worker sisters and have no need to work away, boasting two pairs of wings on their backs. What these are good for, however, does not become apparent until a mild day between June and July in North America, when it is warm enough.

When the moment comes, the whole anthill is seized by a curious unrest, unlike anything seen in the nest. An innate urge forces the flying young queens and males—still around, for once—toward the nest's exits. Anyone else wanting to follow the call is held back by scrappy ant workers. Only on a secret signal do all the nests of a particular species open their doors to release the reproductive insects all at once, and in one fell swoop the exits are swarming with flying males and young queens. They traipse around somewhat aimlessly and eventually take to the air for their nuptial flights.

Queens such as this black garden ant typically only leave the nest for their nuptial flights. Afterward, they focus on laying eggs.

The timing of the nuptial flight during the summer depends on the species of ant. Ants of the species *Temnothorax nylanderi* fly out in the two hours before sundown, while reproductive ants of the species *Temnothorax unifasciatus* prefer to swarm in the morning, around dawn. These different swarming times ensure that nuptial flights do not result in hybrid couplings between males and young queens of different species. Some ants don't leave the nest at all: Young army ant queens stay home on their wedding night, awaiting their lovers in the safety of the nest.

The choreography of nuptial flights also varies according to species. Some males move around in thick swarms of young bachelors, appearing like dark clouds from a distance. If a young queen collides with one of these groups, it swiftly descends into an enormous orgy in full flight. Other queens prefer solid ground. They seek out a romantic spot and emit pheromones, scent signals that no male can resist. Young queens can have sex with as few as one or as many as a dozen males—but just this once. Once the wedding day is over, that's it for life.

This is why the queen diligently gathers as many sperm as she can and stores them inside a special pouch known as the spermatheca. This can hold a few hundred million sperm cells. Not all that many, when we consider that many queens will produce up to 150 million offspring in the next ten, twenty, or thirty years of their lives. A third to a half of these sperms will fertilize an egg and thus contribute to the next generation of ants. This is a considerably better ratio than human sperm enjoy, as only around one in every 250 million human sperm cells succeeds in merging with an egg. Human sperm also have an expiration date of around a month after production, while the cells stored in an ant queen's spermatheca remain viable for decades.

Once the sun begins to sink toward the horizon, the wedding celebrations are over. These millions of future mothers and fathers have little to say to each other after sex—and very different fates await them.

Like most ant queens, the queen of the species Prenolepis imparis *(right) mates with the male during a nuptial flight early in her life.*

FLYING BUNDLES OF SPERM

To cut a long story short: All the males die. Their lives serve a single purpose: transporting sperm cells containing their genetic material to a queen of the same species from another colony. Male ants are little more than flying bundles of sperm and by far the most boring ants in a perfectly organized matriarchal state.

Since the males serve as a sort of sexual delivery service, ants do not invest much in their men. They are not even worth a single sperm from their mothers' nuptial flights. Instead, male ants develop from unfertilized eggs, which the queen only lays if another nuptial flight is imminent, this time for the young queens. At least male ants are not required to work.

They grow, eat until they are big and strong, strike out on their nuptial flights, deliver their sperm, drop dead, and are then carried back to the anthill by whichever ants finds their bodies first, before being promptly gobbled up as handy snacks. And that's the life cycle of a male ant. Of course, there are a couple of interesting exceptions.

Several species of the genus *Cardiocondyla*, tiny ants found around the tropics and subtropics, have devised a second kind of male: one without wings, who is not too keen to die right after his first sexual encounter. These rebels remain in the nest to fight their brothers-cum-rivals. Some species of wingless males also defend a small patch of territory within the nest. When a fight breaks out, it might be one of the fighter males who loses out, but it's worth it: Any male who encounters a young queen on his territory is—with a little luck—able to mate with her, and he may even come out alive afterward. A fighter male can mate more than fifty times and keep producing new sperm, a privilege denied to his flying cousins.

While the primary aim of most male ants is to copulate with a queen during a nuptial flight, fighter males of the genus Cardiocondyla *live for several months and often risk their own lives fighting for access to females.*

In good times, when food is plentiful, young queens prefer sex with these wingless tough guys, avoiding the nuptial flight altogether. Mating in the safety of the nest is less dangerous than a journey into the unknown. They would rather stay home, couple up with one of their brothers and have their mothers declare them co-queens. Eventually, when life in the colony gets a bit cramped due to the number of laying queens, one will round up her entourage and settle down elsewhere in the neighborhood. When the colony finds itself on the breadline, however, young queens will ignore these scrappy couch potatoes. In times of hardship, it's better for them to seek their fortune elsewhere and try their luck finding a better home.

And after mating, luck is the one thing a young queen desperately needs.

THE QUEEN MEETS HER FATE

After fun, work is the next item on any ant queen's agenda, which is why they establish their own colonies following their sexual encounter. It is rarely a good idea to set up shop on the spot where you've mated, so the queen flies a little further away, landing in a new place destined to be her new home . . . or her grave. Most queens demonstrate little flair for choosing their own territory, particularly as the search itself is not easy. Paths, squares, and roads in cities mean there is no loose subsoil into which they can dig, and bodies of water do not provide good landing sites. Any attempt to stop here quickly proves foolish. All kinds of hungry spiders, lizards, toads, and unfamiliar ants might be lying in wait, waiting to grab themselves a tired, tasty morsel. If a queen survives this risky flight, managing not to be drowned, eaten, or trodden on, her flying career soon comes to an end as she breaks or bites off her wings. They will only be a nuisance for what lies ahead.

Her as-yet-unborn new colony needs a home as soon as possible. The queen has three options: She can work hard to build a new nest by herself, she can join an existing community of ants of the same species, or she can become a lodger in a nest belonging to ants of a different species and take control of the foreign colony in a surprise coup. The choice she makes here is a question of the temperaments typical of the different species and the opportunities that present themselves. Either way, every option has its risks.

FIND YOUR DREAM HOME

Building a new nest is certainly the most exhausting option. First, the queen must find a small place of refuge. She is not choosy. Like any true pioneer, she is happy with any sheltered little corner, be it a hollow in a tree or under a rock. If no such place can be found, she must dig a hole herself or gnaw one in some wood. In her little hideaway, known as a claustral cell, the queen is reasonably safe from predators and can focus on producing her first eggs.

Nevertheless, the brave founding queen has the disadvantage of being unable to access the resources and infrastructure of a mature colony. She has no one to look after the brood or to feed her and the larvae. In the early days, she must survive on nutrients from fat reserves built up during her childhood. She also breaks down her useless flight muscles and a part of her brain to feed herself and her first larvae. These reserves are enough for most species of ant, allowing the queen to remain inside her claustral cell for the few weeks or months of the founding phase, which is why we refer to this as claustral colony foundation. Queens of other species, however, have no appreciable dowry and are forced to leave their hideaways from time to time to find food, in what is known as semi-claustral foundation. All these supply corridors are dangerous: New mothers can come to all sorts of harm outside, or a hungry contemporary may discover the helpless larvae and gobble them up. The queen is happy to do this herself when supplies

The first worker (left) in this lab-based colony of Temnothorax longispinosus *has already hatched and is helping the queen (right) to establish the nest.*

are low. She will sacrifice a couple of her eggs or larvae to feed the others. The larvae will also nibble at one another on occasion. When it comes to cannibalism, ants have a very relaxed outlook.

If the supplies or supply runs hold out, the queen produces a few unusually small workers, known as "minims." From this point onward, responsibility for the colony lies in their tiny mandibles while the queen returns to her life's work of egg-laying. Minims really try their utmost, but they are also extremely cautious, almost anxious in their undertakings. Unlike their subsequent siblings, they can't display much bravado because there are so few of them and every worker matters if the colony is to survive.

It's far less work—and much less risky—when a newly founded colony is formed by several young queens joining together to create a community, rather than by a single lone queen. Teams of this kind function surprisingly well, at first.

However—as is often the case with housemates—once everything is going according to plan and a sense of routine sets in, there is greater risk of the system breaking down. Once the first workers are in position, the queens begin to get tetchy and start attacking one another. Only in the rarest of cases do cocreated communities such as these develop into colonies led by several queens who are not related to one another. This is referred to in the field as polygyny. Nevertheless, the founding queens almost always become embroiled in fights because each of them wants to be the sole queen in a monogynous colony. And it's in their interest to win such fights since the losers often pay with considerable injury, and if they don't die from their wounds, workers and even their own daughters will be happy to finish them off. The queen is dead! Long live the queen!

MOM'S HOTEL

Faced with such odds, the young queens of many species plump for a stay at Mom's Hotel, returning to their home nests directly after the nuptial flight. When it comes to seeking out a new love affair, the red wood ant *Formica rufa* never strays too far from the anthill where she was raised. So, when the time comes to look for shelter, they often find themselves barging in on their worker sisters, who are busy with other activities. The workers recognize their sisters and carry them back to the nest. Sometimes the young queens are lucky enough to get adopted by the workers of another anthill instead of killed. This, however, only works for species with polygynous colonies, which have multiple queens. If the new queen is accepted, she has an easy life ahead of her. She needs only concern herself with egg-laying; the workers take charge of all other business within the nest. The selfless workers are the only ones disadvantaged by this arrangement: The queens in the colony are not all their mothers and the workers are not all their sisters, and this poses a problem in terms of evolutionary biology. Many are still closely related, but some are simply not related at all. Mom's Hotel often turns out to be Mom's Village.

At some point, the village will threaten to burst at the seams. A large number of queens produce an even larger number of workers, who need more and more space and, above all, more food. Their immediate surroundings are soon munched bare and the distances the ants must travel to find food grow longer. At this point, it's time for some of the queens and a portion of the colony to move out and form a new branch some distance away. Since none of them can fly, they travel to their new home on foot. This new home is closer to the main hill and subject to much greater competition than if the queen had flown out right after the nuptial flight, but she is able to take an entire household with her, including workers of different ages, larvae, and eggs. And if she finds she is missing something, it's not the end of the world. Daughter colonies are often closely tied to the mother nest so that workers can come and go from each nest without running into trouble. Together, the nests form what is known as a supercolony. This is why red wood ants mostly live in communities with several ant hills located in close proximity to one another. The forest edge lies in the hands of a veritable dynasty.

Compared with individual nests with a single queen, the great advantage of polygynous communities and supercolonies is that they make the colony practically immortal. When one queen's life draws to an end, there are always others to take her place. If one of the nests is destroyed, the surviving workers and queens can find new homes in the other branches of the colony. By contrast, an individual nest with a single queen will collapse once the queen dies. Once again, ants truly take their strength from their community.

¡SU CASA ES MI CASA!

But what is a queen to do if she's recently mated and wants to avoid the effort and unpredictability of establishing a new nest, but can't find her way back to her old one? Wood ants and other ant species have developed a risky and particularly heinous strategy for such circumstances: capturing the nests and workers of other species. The young queen must camouflage herself during this hostile takeover, entering the nest undercover to ensure she is not mistaken for prey and gobbled up. To do this, she steals the target colony's scent chemicals, taking them from nest materials or host workers, which essentially give her a free pass into the colony. She slips into the anthill and tries to force her way into the central chamber where the queen lives. This is followed by a showdown worthy of the Wild West, and it can go several ways. The insurgent queen often kills the rightful queen, but sometimes she only deposes her predecessor and leaves it up to one of the queen's own daughters to kill their mother. Occasionally, however, the reigning queen is faster and stabs her rival or throws her to her workers as a snack. Either way, it's certainly a bloody affair.

At least, *most of the time.* Species of the genus *Lasius* practice a more subtle form of hijacking. They seek out nests without queens, such as those where the queen has died of old age. The shiny black *Lasius fuliginosus* or jet ant will assume control of colonies belonging to the yellow *Lasius umbratus*, whose queens like to capture the nests of the black garden ant *Lasius niger* or the brown tree ant *Lasius brunneus*. Yes, it gets a little confusing—particularly during the transition period when workers of both species live and work inside the nest side by side.

The new queen's long-term goal is to establish a colony of her own. The workers of the host species serve only to bridge the difficult initial phase by caring for the insurgent queen's larvae and providing food. Gradually, these host workers die off and are replaced by the new queen's own workers. During a transition phase you will encounter mixed nests, often with the older host workers serving to collect food and the younger, insurgent workers employed inside the nest. Scientists refer to this phenomenon as "temporary social parasitism." It is "temporary" because the victims are only temporarily exploited (since they are all dead by the end of it) and "social parasitism" because the parasites ruthlessly exploit the social behaviors inherent to ants.

Whichever path a queen takes to establish a colony of her own, eventually—we assume—when the excitement dies down, the queen is surrounded by her daughters and is finally ready to rule. . . .

Or is she?

CHAPTER 3

EFFECTIVE ANARCHY

Army ants of the species Leptogenys distinguenda *move home every few months, carrying with them their larvae and pupae (left) as well as the occasional guest, such as this snail.*

If any large group is to function, someone has to have the final say. At school, it's the teacher; at work, it's the boss; and in a nation, it's the president, the prime minister, or chancellor. When humans began showing an interest in the lives of ants, kings were still at the top of this hierarchy. Their rule was absolute, their word was law, and their subjects built cities, harvested crops, and waged wars all in accordance with their whims. Exactly—as early researchers were thrilled to discover—like ants do when ruled by a queen. What a wonderful analogy, what a fantastic proof of the God-given nature of monarchy! . . . And what a huge mistake.

At first glance, of course, there are obvious parallels between people and animals. And it is truly astounding to see how capable ants are. They live in communities, some of which comprise more citizens than countries like Denmark. They provide sufficient food for everyone, clear away waste, and regulate the temperature within their nests, making sure no one gets too hot or too cold. To achieve the same things, humans need someone to take charge, give out commands, and keep a close eye on proceedings. We need leaders, whether we call them chiefs, mayors, kings, or emperors.

Ants, on the other hand, do it all without a leader in sight.

Their so-called queen has no real duties within the colony. She is essentially a stay-at-home mom, the colony's egg-laying machine. After all, that's what a queen does, day in, day out: She eats, and she lays eggs. From the moment she establishes her colony, her life is as dull as it gets. It's her many daughters, the workers, who have the real say. And so, the form of government seen inside an ant colony is less reminiscent of absolute monarchy than it is of democracy—and anarchy is always a threat underneath the surface.

WANTED: 1,000-BED PROPERTY WITH KITCHEN EN SUITE

To examine how ants make decisions, we are going to take a closer look at the biggest logistical challenge that any ant colony must master: relocating the colony.

There are many reasons why an old nest may no longer serve its purpose. Over time, colonies living in small twigs or acorns on the forest floor may find their nest chambers collapsing of their own accord or accidentally trampled by a deer or bear. Often, the community will simply have become too great in number, requiring more expansive accommodation. They may also have munched their way through all nearby food sources and move on to seek more elsewhere, as is often the case for army ants. Perhaps an aggressive neighbor has just moved in and it makes sense to get out of the way. Sometimes, it's the fault of nasty researchers who dig up the nest and take part of the colony back to their laboratories. Whatever triggers the move, the colony faces a huge project, which, depending on its size, can be akin to relocating a megacity as big as Tokyo. Before embarking on this mammoth task, the colony must first take an urgent look at the local property market.

A handful of prospective buyers—scouts—move out, checking every available nook and cranny to determine whether one would be suitable for their colony. Their list of requirements is a long one. First, the new home must be accessible on foot because ant workers cannot fly. It must be large enough for the entire colony and provide protection from predators and bad weather; it must be made of materials that allow walls to be demolished and reinstalled; it must offer pleasant levels of moisture, even during the dry season, but not become dripping wet when it rains; it must keep out the cold in winter without being baking hot in summer—and so on and so forth.

It's quite a stroke of luck when said scouts happen upon a suitable location. Before making a decision, they carry out extensive checks. They run in and out of the potential nest several times, examining every square inch of the space. Inspections of this kind can last up to an hour and only afterward do the ants decide. After all, it's up to them to decide. Though these little scouts are just some of dozens, hundreds, thousands, or maybe millions of ants, it all comes down to their instinct, their experience, and perhaps even their taste. Nobody is whispering in their ear, telling them what to do next. No real estate agent, no sisters, and certainly no queen.

It's a lot for a little ant's brain to handle.

SIX-LEGGED COMPUTERS

Yes, ants do have brains. They are not especially large and, depending on species, may be smaller than forty thousandths of an inch (1 cm³), weigh just a few millionths of a gram, and get by with around 250,000 nerve cells. But the real problem is trying to dissect them. How do we do it? With a razor blade, delicate instruments, and near-infinite patience.

First, you break the razor blade in two and attach one half to a scalpel. Then you decapitate the ant —a pleasant death, if we are to believe Dr Joseph-Ignace Guillotin, a man who did not invent the instrument of capital punishment named after him, but who recommended its introduction during the French Revolution on "humanitarian grounds," saying, "The victim will feel little more than a refreshing cold sensation." An ant's head is too small to hold or move, so you place it in liquid wax, allowing this to solidify in a dissection tray on ice.

Now for the tricky part. You make three cuts through the ant's hard exoskeleton, known as the cuticle. A sort of flap will appear that you must open carefully, like a window. You will now be able to see the brain. Scientifically, an ant's brain is referred to as the supraesophageal ganglion due to its position in the upper part of the oral cavity and—despite its diminutive size—is split across several different centers, each performing specific tasks. Signals from the antennae are picked up in the antennal and dorsal lobes, new images from the eyes are received in the optic lobes. Things get really interesting in the central bodies and particularly in the mushroom bodies, so named because under the microscope they look like mushrooms with hats. This is where the ant's brain processes incoming information, storing experiences that will be of use to its owner over the course of her life. You don't have much time at all to get a good look at everything because RNA molecules are sensitive and disintegrate if you don't work fast enough, meaning all your work will have been for nothing. Swap your razor blade scalpel for a pair of ultrafine tweezers, carefully remove the brain from the surrounding tissue, and lift it out of the head.

And voilà! You have extracted your first ant brain. Now all you need to do is repeat the process a few dozen more times until there's enough material for your investigation.

THE WISDOM OF GENES

The size of an ant's brain not only depends on the ant's species and caste, it boasts a fascinating link to the ant's principal job. Contrary to what you might expect, queens' brains are not all that big. Once the nuptial flight is over, they have established their own nests and the first workers have set to work, queens experience a radical form of mental degradation. All the areas that are no longer needed—since the queen's sole focus now is egg-laying—shrink. This includes flight control, as well as the center for foraging and caring for the brood. So, it's not entirely true to say that the queen no longer *needs* to work—in reality, she simply can't. She is now too dumb to do so.

For her workers, it's a different story. The workers continue to accumulate experiences throughout their lives, accruing so much experience that, in some species, it leaves an anatomical trace. The mushroom bodies in the brains of young workers in the hatchery are still very small, because they have known nothing other than their queen, her eggs, larvae and pupae, and the inside of the nest. Experienced workers employed outside the nest, however, are constantly having to memorize paths to valuable food sources, and consequently have much larger mushroom bodies.

The anatomical differences are not enough for us to understand why an ant does what it does. Instead, we have to look at what the ant was doing shortly before her untimely death and which genes were active when she was decapitated. Each individual worker possesses all the genetic material necessary for a whole host of tasks, but she uses only the portion that she requires at the given time. She makes copies of this in the form of RNA molecules, which we then isolate using biochemical methods, enabling us to identify them by their structure. Comparing workers tasked with a variety of jobs with their accompanying RNA mixtures tells us which genes are responsible for which specific behaviors. This method allows us to demonstrate that workers employed to take care of the brood exhibit a particularly active gene for the protein *Vitellogenin-like A*. If we restrict this activity in living ants, they suddenly lose interest in larvae and pupae and begin instead to care for the adult ants with which they share a nest. In nature, scents control gene activity, and it appears that the *Vitellogenin-like A* gene is important when it comes to allotting tasks.

YOU ARE ALL INDIVIDUALS!

The combination of differently interconnected nerve cells in the mushroom bodies and other brain centers with different gene activities gives rise to an unexpected quality that few would believe ants to possess: They all have different personalities.

After a particularly grueling period of study, Evelien Jonge-pier, a doctoral student of mine, was able to demonstrate pre-cisely this. She "interviewed" 3,842 workers from 102 colonies of the species *Temnothorax longispinosus* from upstate New York to ascertain their professional preferences. To do this, she placed each insect one by one in front of either a pupa in need of help—to draw out broodcare behavior—or dead ants from a foreign colony, to provoke a defensive response. If ants were the mind-less robots we humans sometimes believe them to be, every worker would have reached out mechanically for either stimu-lus, tending to the pupa or attacking the insurgent ant. But the ants did not all respond according to Evelien's predictions. Some did not even think about being roped into specific tasks. A few paid particular attention to the pupa but were not prepared to be drawn into a fight with the foreign ant worker. Others lunged at their enemy, paying no heed at all to the helpless pupa. Another subgroup ignored both the pupa and the enemy ant and simply did nothing. In all probability, different genetic activities prede-termined which stimulus the ants would respond to.

The significance of personal experience is demonstrated by observing workers of the clonal raider ant *Ooceraea biroi.* They are predatory creatures, feeding principally on the lar-vae of other species of ant, having forced their way into their nests. In the lab, a group of workers prevented from preying on others will eventually give up and forgo its raid altogether. Instead, the ants will focus on caring for the brood, whereas the control group whose attempts to hunt are not sabotaged feels emboldened by its successes, obtaining more food. So, even ants feel frustrated with their lack of success from time to time and choose to switch jobs. What's clear, though, is that ants are not just unthinking six-legged robots. They nurture individual interests and dislikes.

But we can't go so far as to suggest that ants have free will. While each worker may have developed her own personality, no ant is pondering the meaning of life or reflecting on whether she should spend the day foraging for food or making improvements to the nest. The differences in their behavior stem from the biological mechanisms described above, which largely determine an ant's character.

LET'S MOVE HOUSE

When it comes to deciding whether a site is suitable for a new nest, the little scout we met at the beginning of the chapter relies on the genetic specifications of her caste and her age as well as her own personal experience. If she concludes the site doesn't match her vision, she leaves and seeks out another, better place to stay. But if she thinks it could make a good new home for her colony, she runs straight back to the nest to get a second opinion.

It's certainly remarkable that an individual worker is tasked with answering such a difficult question, and one that affects the fortunes of the entire colony. It doesn't mean that all her sisters will be of the same opinion, however. Another scout might have discovered another suitable place in the meantime. So, before the decision for or against the new site is made, several experts are sent in to check it. The scouts use scent trails, personal invitations of sorts, to draw their colleagues toward the site of their choice. Once there, each ant will inspect the site as thoroughly as if she had discovered it herself, making her own judgment. If she likes it, she will recruit other experts so that an attractive abode soon has more and more devotees. Scouts do not generally inspect every potential nesting place. In fact, the vast majority only visit one site, and only a small percentage compare two or more sites. But as soon as the number of enthusiastic fans exceeds a certain level, the decision is made by committee for the entire colony: This must be the place! The number of votes required depends on how pressing the move is. If the

Not every worker knows where the new nest is, so her sisters carry her to her new home—just like this Sahara Desert ant, Cataglyphis bicolor.

old nest has been destroyed or if hostile ants are threatening the queen and the brood, the ants will leave the decision to just a couple of workers. If, however, there's no rush, they will be pickier, comparing several potential nesting sites before they ultimately decide.

This is when the ants' work begins in earnest. They're in the freight business, transporting everything of value from the old nest to their new home piece by piece. Using their powerful mouth parts, they transport not only eggs, larvae, and pupae, but also their younger sisters (whose gene activity does not allow for any active role in the move) all the way to their future home. Ants of certain species that are transported in this way adopt a specific position for transport, tucking in their abdomens, antennae, and legs, and making themselves porta-ble like little suitcases. Species with small colonies make the move in one go. Larger colonies living in more complex nests, like leafcutter ants, require several weeks for the move, which sees a flurry of activity on their specially created trails. The first workers to arrive at the new address set to work furnishing the place. They dig tunnels and chambers, ensuring that the nest is reasonably prepared for the gang to move in. The queen her-self will often then make her way to her future home on foot, flanked by a powerful bodyguard. This is when she sees it for the first time. She has no say in choosing or furnishing her new home. Her daughters are the ones who make the decisions.

ELECTIVE AFFINITIES

The move to the new nest is just one example of the many ac-tivities in which the queen is at best a passive participant. De-spite her title, she gives no commands and has no sovereign power to exercise. It is her workers who make and implement decisions both big and small—in a manner that is decen-tralized, democratic, and a little anarchic, but almost always highly effective. While they work both inside and outside, the queen simply hunkers down in her chamber, with little idea of what might be happening in her colony.

But it gets worse. It's not just that the queen is a nobody in her own kingdom; she's under her workers' control, too. This is particularly noticeable in colonies with a sole queen who has mated with only one male. To understand this, we have to look at heredity. Unfortunately, this gets a bit complicated, but I'll try to keep my explanation as simple as possible. If you aren't interested in the details, feel free to jump ahead to the penultimate paragraph in this section and read on from there. There won't be a test at the end of the book!

Female ants develop from fertilized eggs and have two par-ents: their mother, the queen, and their father, a male. Each parent has bequeathed to its child a complete set of genes, meaning that females possess a double set of genes—in scientific terms, a female is described as "diploid" for this reason. There are different variations of many of these genes, known as alleles. Imagine it like a kitchen mixer, Model A: 100W, Model B: 250W, or Model C: 500W. The mixer stands for the gene, the model stands for the different alleles. The function remains the same but there are slight differences, which may emerge during use. Since the queen is pretty set for kitchen gadgets, since she has every gene in duplicate, she has different alleles for many genes. Let's say she has alleles A and B for one particular gene.

Now the queen does what she does best: She produces an egg. The egg inherits one sample of each gene and only one of each of the two alleles. Eggs only carry either the A allele or the B allele. But the eggs have not been fertilized yet. If a male is to hatch out of one of the eggs, it doesn't matter because all that's required is an unfertilized egg with a simple, or "haploid," gene set. But most of the eggs are fertilized, fusing with a sperm cell from the queen's sperm pouch. This sperm contributes an allele from a male who is foreign to the nest, making this Model C. The fusion of egg cells and sperm can produce two different outcomes: AC or BC, where A and B are from the queen and C is from the father. So, each daughter is 50 percent related to her mother. But this doesn't affect just one gene, it affects all of them.

This means that the degree of kinship between two sisters has two extreme options: Both insects can inherit different alleles for all genes from their mother, making them genetically 50 percent identical because the alleles from their father are the same. Alternatively, they inherit the same allele for all the genes inherited from their mother, making them 100 percent identical. Almost all sisters lie somewhere in the middle, giving an average rate of genetic kinship of 75 percent, which makes them more closely related to each other than they are to their mothers.

But what about sons and brothers? Since the eggs that produce them are not fertilized, male ants inherit all their genes, and all their alleles, from their mothers—they are mama's boys through and through. And this remains true whether the male inherits the A allele or the B allele. Unlike their sisters, however, another result is possible. Males can't expect genetic material from a father—as in the example of the C allele—because they have no fathers. This means that a male ant may share up to 50 percent of his alleles with his sister, ensuring a maximum 50 percent genetic kinship, and brothers and sisters may also share no alleles at all, making their degree of genetic kinship 0 percent despite sharing the same mother. On average, brothers and sisters tend to exhibit a degree of genetic kinship of 25 percent.

Of course, ants don't calculate their degree of kinship analytically. And the asymmetrical nature of the kinship between sisters (75 percent) and between workers and their brothers (25 percent) would not be too much of a problem were it not for the mechanics of kin selection that has developed over the course of evolution. This theory states that insects offer greater self-sacrificial support according to their degree of kinship, as this ensures that the alleles they share are passed on to the next generation. Bad luck for the brothers! While the queen would prefer a balanced ratio of daughters to sons, the workers of many ant species favor their sisters to such a degree that they have no qualms about killing "male" eggs and feeding them to female larvae—a palace revolution in the nesting chamber. The sisterhood has no time for fraternity.

The queen is firmly under her colony's thumb, even when it comes to her specialty: egg-laying. In compensation, she receives the best food and uses special scent signals, pheromones, to prevent her worker daughters from producing their own offspring. In most ant species, this monopoly is a privilege reserved for the queen alone.

COMMUNITY MATTERS

But why should ants care how closely related they are to the others with whom they share a nest? Modern biology can now answer this question. As life has evolved over billions of years, mechanisms have developed to ensure that improvements are passed on to subsequent generations. This might be a bacterium's ability to digest a new sugar, enabling it to grow faster and divide more frequently. A tree with deeper roots might be able to reach groundwater, allowing it to thrive in areas where there has long been competition for water. Or perhaps it might be a bird of paradise, which uses its colorful feathers to impress more females than its rivals, enabling it to assemble a large harem. Genes lie at the heart of all these successes, governing metabolism, growth, appearance, and behavior. If a gene can induce an organism to reproduce, it will be passed on, giving it new opportunities in the next generation. A gene that fails to do this will disappear without further ado. Ultimately, it all depends on whether genes can make the leap from one generation to the next. But it is exactly this which presents a problem for ant workers: They have no offspring of their own.

Workers compensate for their lack of fertility with another strategy: Like some bees and wasps, and all termites and naked mole rats, ants are eusocial. Scientists ascribe this label to species in which most animals eschew producing their own offspring, instead tending to the offspring of a closely related female, the queen. Lions are not eusocial because each of the females in the pride is fertile. Young moorhens, hatched the previous year, sometimes help to raise their siblings born the year after, but only one other generation ever takes part.

The sterile ant workers place their faith in their queenly sisters. As we have seen, they share 75 percent of their genes with the young queens, as they do with all their sisters—more than they would with their own daughters. By putting all their strength into helping the young queens establish a new colony, these workers guarantee their genes are passed on to the next generation. And this works best within a strong community.

Every worker can rely on her sisters to look after the offspring if she dies, so she is able to place herself fully in the service of the colony, regardless of the consequences. She will square up to the most fearsome foe without a moment's hesitation, risk everything to find food, and even offer herself up willingly as food for her sisters if the colony is starving. Since the colony persists for generations, it is able to create a well-protected nest that one ant could never build alone. Though each ant is small and weak on its own, the colony is strong and able to defend itself. By contrast, a solitary wasp must build a small new nest every year for its few eggs and must leave it unprotected when it goes out to seek food. It has to fight off attackers all on its own. If the wasp dies, her brood will also die, and all her effort will have been for nothing. Being eusocial has enormous advantages if you are happy to go without sex.

So how did ants come up with this unusual and ingenious strategy?

SHORTCUTS TO MONARCHY

The path to eusociality was a slow and subtle one. The ancestors of the ants were still struggling to get by and raise their broods alone, but at some point, a couple of genes changed, changing their behavior and, later, their physical makeup, too. We still do not know exactly what happened, but research on different species of wasps and bees—ants' flying cousins—has allowed us to directly observe some of the developmental stages of sociality. And it is here where we found two pathways.

One of these paths begins with several fertile sisters who each established their own families within a community. Each mother was responsible for tending to and defending her brood. The sisters did not share these responsibilities until the second stage of development. At this stage, the responsibility fell on several shoulders. However, since there were only a few adult insects, the loss of one of them posed a huge risk to the entire brood. This risk shrank when one of the females became so dominant that she was able to prevent the others from producing offspring. The subservient sisters lost the ability to reproduce but retained their maternal instinct and dedicated themselves to raising their young nieces and nephews. The generations ultimately remained together in the same nest permanently, raising the queen's offspring—as a eusocial community.

The second path to eusociality begins with a single mother and her first generation of daughters. This generation requires the mother to persuade her children to remain in the nest and help around the house, including caring for the next generation's brood. This kind of arrangement would also have had advantages for the first-born daughters. Their mother had to cope alone, so she will not have been able to tend to this generation particularly well. As a result, these daughters were small and weak and less likely to successfully establish their own nests. By joining Team Mom they could at least give their better-prepared sisters—who were very closely related to them—a fighting chance.

In both scenarios, the workers' bodies increasingly adapted to the new conditions because the more successfully a colony was able to establish itself and multiply, the more consistently the workers were able to adapt to their new lifestyle. Although each worker has the same genes as the queen to this day, many of these genes cease to be required by the workers and become inactive. A worker, for example, does not need wings because it will not be required to leave the nest and look for a mate. Thus, the genes for wings will remain inactive throughout a worker's life, even though they lie dormant in every single one of her cells. By contrast, the genes that govern caring for the brood must be switched on earlier than they would for solitary insects, which are usually concerned first with finding food before they produce offspring and have to tend to them. A wealth of genetic switches must be flipped or adjusted. Nature spent several dozen million years experimenting, rejecting incorrect attempts, and promoting good ones until eventually, at least 100 million years ago, the first ants—as we recognize them today—crawled about between the toes of the dinosaurs. In the meantime, over 16,000 species, each with their own special characteristics, have developed out of these. A plethora of bustling colonies with helpless queens and a multitude of anarcho-democratic workers.

CHAPTER 4

COMMUNICATIVE SENSUALITY

———

Two workers of the species Formica obscuripes *exchange liquid food and chemical messages.*

L et's be honest: When we observe just one species and one caste of ants—such as the workers of the western thatching ant, *Formica obscuripes*—it's pretty hard to tell them apart. We paint little colored spots on their cuticles so that we can tell the difference between Ant 4278 from Nest A and Ant 3321 from Nest B. And it's often practically impossible to precisely determine the species. Some species of *Dolichoderinae* ants of the genus *Tapinoma* can only be told apart by observing the makeup of the male's genitals. Just try getting a look at them! Males are not especially numerous, either. To us humans, they often look the same.

But not to ants.

Ants can easily tell members of their own colony from members of a different colony or species—often with fatal accuracy, as you will discover. They are also able to discern whether an ant is responsible for looking after the larvae in the nursery, keeping the nest in good condition, or foraging for food outside the nest. Queens of the neotropical species *Neoponera inversa*, who establish a nest together, are even able to recognize themselves as individuals. However, I would suggest that this degree of personal uniqueness is not present in larger colonies. Even if individual insects differ behaviorally and display some degree of individuality—as we saw in chapter 3, Effective Anarchy—the differences between those caring for the brood, those gathering food, and those guarding the nest are much larger, such that ants view themselves more as part of a distinct group and less as individuals. For example, those caring for the brood inform those gathering food that food is required and accept prey from them in order to feed the larvae. None of the sisters is the slightest bit interested in whether Ant 5241 slept badly or had a close shave with a lizard that day.

Fundamentally, however, ants are very adept at sharing information, be it about the route to a newly discovered herd of aphids, the existence of unexpected prey, or a raid by another hostile colony. Information about internal affairs is also exchanged, ensuring that every worker knows whether the queen is still fertile and hard at work producing eggs. Gossip about the goings-on in the royal family is not the sole preserve of humans.

However, not all species of ants share this kind of information with one another. The workers of some primitive species of ants are of a more solitary nature and somewhat taciturn. The workers of highly developed species, by contrast, are often well-informed and information seems to spread throughout the colony as if by magic.

Still, we have succeeded in deciphering some of the tricks ants use to communicate with one another.

SEEING THE LIGHT

As far as ants are concerned, looks don't matter. This is perhaps unsurprising for insects who live out their lives in fixed, uniform exoskeletons, making them as capable of gestures and expressions as a knight in full armor with his visor down. Ants have two highly developed compound eyes on the sides of their heads, and queens and males also possess three smaller, upturned simple eyes called ocelli. Ocelli are light-sensitive, react to UV light, and are even able to detect the direction from which sunlight is polarized. We will deal with polarization in more detail in a later chapter, but for now it suffices to say that this acts as a kind of celestial compass, telling you where the sun is, even on a cloudy day. Queens and males use their simple eyes for navigation and flight stabilization on their nuptial flights.

The job of the compound eye is to perceive visual images. You're probably familiar with the larger versions seen on flies, bees, and dragonflies. They are a little like an upturned sieve, but each hole in a compound eye constitutes the lens of an individual rod-shaped eye. This individual eye makes up a specific point of an overall image, which the compound eye sees. But don't expect great quality. Depending on type, a compound eye comprises between three and thirteen hundred individual eyes, so the resolution is terrible. For comparison, even a cheap cell phone camera captures images across five million pixels. What an ant sees is more akin to a rough newspaper image. And as if that weren't bad enough, it's also blurry because the lenses in the individual eyes are rigid. It's in color, at least, though an image as perceived by an ant's eye would appear quite bluish to us humans. Like bees, ants perceive the color red as mainly black, but to compensate for this, they are able to see UV light. Nevertheless, it's quite difficult for a worker's standard compound eye to differentiate between its sister and a pebble of the same size. And in an ant's eyes, there's not much more to see—especially when you consider how short their legs are.

This doesn't bother the ants. They have no intention of mapping out their surroundings in HD or 4K, and place greater value on other characteristics, using their eyes as highly sensitive motion sensors. And that's all that's required to quickly ascertain that a shadow has suddenly disappeared from one pixel, before popping up in another. In flies, this process occurs so quickly that a TV show would seem to them like a series of stills, while we think we are seeing fluid movement. When it comes to matters of speed, compound eyes are clearly the better design. Species of ants with very large eyes are particularly adept at using their optical motion sensors to identify prey and pounce on the specimen under their noses. Workers of the Australian species of bulldog ants, of the genus *Myrmecia*, will even follow a finger moving back and forth. As a researcher of ants, I have always found this a somewhat uncanny experience, particularly since I know bulldog ants are very capable of jumping and can have a nasty sting.

The Australian bulldog ant has excellent eyesight and can follow the movement of a finger with its head.

Flying male ants use sight, among other things, to find the young queen of their dreams during their nuptial flights. Compound eyes also make excellent celestial compasses. In short, they are just as good as our eyes; it simply depends on your priorities.

However ants recognize one another, we can be certain they do not rely on their sense of sight.

A SIX-LEGGED CHEMICAL PLANT

The secret is scent. Just as we might laugh at the ants' poor eyesight, they would turn up their noses at our miserable sense of smell. Only figuratively, of course, because ants do not actually possess noses as such. They smell and taste mainly via their antennae, or feelers. These are covered in clusters of ultrasensitive chemical sensors, which react to different substances. Alongside practical smells, such as sweet or floral, which we ourselves can pick up, ants also have a nose for signals that escape us entirely. They can sense carbon dioxide, which helps them to keep their underground nests properly ventilated before the entire queen-dom suffocates. They also have a sensor for moisture, which prevents immobile eggs, larvae, and pupae from rotting or dry-ing out, having been moved to a better position when required.

The most surprising thing is not what ants are able to smell in their environments, but the scents they themselves produce and use: Ants have developed their very own chemical language.

Linguists may raise their eyebrows at the sentence above, and the ants' repertoire does not extend to *Hamlet* or *Emma*, but by using their pheromones, ants are able to send messages like, "Watch out! We're being attacked!"; "I know where there is food"; "I am part of your colony"; "This is our territory"; "I work in the brood chamber"; and "I am a young queen, look-ing for a male for a bit of you-know-what." In short, ants can communicate anything that crosses their minds.

To make sure they really get the message across, ant bod-ies are equipped with countless glands, transforming them into walking chemical plants. These glands are everywhere, in their heads, their thoraxes, and their abdomens, and several of them often produce toxins, digestive secretions, and hor-mones, as well as a cocktail of different messenger substances.

For us humans, chemical communication between ants is almost fully encrypted—but only almost. With the right equipment and a good measure of finesse, we are discovering ever more precisely what each mixture of substances means.

To do this, we shoot the messenger—well, we freeze her. We then place the ant on a slide, which serves as an operating table. As each part of an ant is incredibly tiny, the following steps re-quire a stereomicroscope and tremendously steady hands. Us-ing superfine metal tweezers, we gently pull out the ant's sting-er, along with the attached glands. This has the same effect as it does on a honeybee that foolishly stings someone, ripping outs its innards as it tries to pull its stinger out of the human's skin. If the glands are intact, we place them in a drop of water and separate them. We are not particularly interested in the round poison gland, but the elongated Dufour's gland conceals a wealth of signal substances. We can either chemically analyze their contents or investigate their effects more directly by smearing them on a couple of test ants. Will the workers accept their perfumed sisters, or will they attack them angrily and throw them out of the nest? Perhaps they will simply help them to wash off the secretions? Or will they respond entirely unexpectedly?

As a matter of fact, the substances produced by the Dufour's gland send the ants into a frenzy, akin to placing a "Kick me!" sign on the backs of the unfortunate test ants. The sisters can be relied upon to attack the insects bearing the scent, so much so that some species use these secretions for their own criminal ends. These include socially parasitic ants, which spread the secretions when attacking other colonies, kidnapping and enslaving their larvae and workers. Confused by the secretions, the victims do not attack the insurgents, instead mobbing members of their own nest. The queens of the genus *Leptothorax* are more focused but no less cold-blooded. They mark their rivals with secretions from the Dufour's gland, causing the workers to kill the rival queens. Whatever those ants have got in their glands, it must be strong stuff.

During our chemical analysis, we add the secretion to a solvent and use a gas chromatograph to separate it into its constituent parts. The sample is slowly heated, and the individual components steadily vaporize. A fine stream of helium gas carries them through a hollow tube where the different substances are slowed down depending on their chemical characteristics. At the end, they emerge neatly separated from one another and move into a mass spectrometer, which not only tells us which substance is which, it also tells us what quantity of the substance is present in the secretion.

Ants carry around only a millionth of a gram in their glands, but it is an impressive chemical kit all the same.

DO YOU SPEAK CHEMISTRY?

The "words" that make up the language of ants derive almost entirely from the principle building blocks of organic chemistry. Alongside hydrocarbons, so named because they are composed solely of atoms from the elements carbon and hydrogen, we also find alcohols, aldehydes, esters, and terpenoids, all of which contain one or more oxygen atoms, as well as compounds with sulfur or nitrogen atoms. Sometimes the molecules are elongated, sometimes they are branched, and sometimes they form a ring. Not forgetting formic acid, of course, which serves as a means of communication in small doses.

In order to say something—form a "sentence," as it were—ants often mix substances in different concentrations. They can create an astronomical number of possible combinations out of just a handful of raw materials, allowing them to say far more than a tiny ant brain is capable of thinking. This makes the job of deciphering their language even harder for us researchers.

The nest has its own scent, which each ant carries on its cuticle. This is generally a mixture of thirty to sixty different chemical compounds in exact ratios. To crack the code, we place a dead ant in solvent and analyze the mixture as described above, using a gas chromatograph and mass spectrometer. We are then able to see on our computer screens what ants smell when they encounter the test worker, and we are able to say which colony the ant has come from—the one difference being that we take much longer, require expensive laboratory equipment, and are still only about as accurate as an ant with a slight cold. But at least we are getting a sniff of another world.

Chemicals have several advantages over spoken or visual languages for somebody who is small and lives in the dark. Since pheromones evaporate and are disseminated through the air, they can carry their messages over comparably long distances and reach a large number of fellow ants. If speed is required, light, ephemeral substances are the best option. A lasting signal will be composed of substances that cannot be easily erased. Both varieties are good for sniffing out in the darkness of the nest, where even the biggest eyes struggle to see. A chemical language is the perfect choice for ants. And it's incredible how much they are able to interpret through scent.

THE CHEMICAL CODE

Perhaps the most important message that an ant can communicate is, "I am from the same colony as you." If this message is not true, trouble may arise. The deeper an unfamiliar ant travels into a nest, be it accidentally or on purpose, the more violently she will be attacked. If the insurgent ant roams through the outlying suburbs of a territory, sometimes its inhabitants will only go so far as to scare her away with angry, threatening gestures—a real stroke of luck. But if she is discovered in the heart of their territory, the ants will expel her roughly from their lands by grabbing her with their mandibles and carrying her to the border. If, however, she finds her way into the cramped nesting area, there will be no mercy. A whole horde of ants will descend on their foe, biting or stinging her to death. The difference between friend and foe is purely a matter of smell. Ants of every colony have a distinct nest scent that each ant carries on its cuticle.

Ants usually take sole responsibility for creating their own olfactory ID cards. They produce hydrocarbons, which reach the cuticle via a bloodlike fluid called hemolymph, in glands and specialized cells known as oenocytes. They also make use of fragrant substances from their surroundings, such as the tree where the colony lives. The ants clean and lick one another, distributing the scents over each other's bodies, mixing substances together to create a common scent pattern. An important reservoir and mixer for these scent chemicals is the postpharyngeal gland, located near the mouth. In some species, the social stomach plays a part in the transfer of pheromones when ants feed one another with regurgitated food. These different mechanisms ensure that all the ants in a hill carry a specific chemical profile, which serves as a reliable form of identification. Occasionally, an ant will have the misfortune to participate in one of our studies, in which we wash it thoroughly with solvent to remove every trace of scent. If we then send it back to its sisters, it will be met with suspicious sniffs before being lathered up with the scent of the nest once more. The right scent is as important as the right team uniform.

An ant learns shortly after hatching which team she belongs to. At first, she might feel she could belong to any team. Over the course of her first few days, however, she gradually comes to memorize the scent of her own nest, all while smelling more strongly of it herself.

NOSE YOUR PLACE

In the pitch-black of the nest, however, ants not only pick up the scent of their own colony, they can also smell which caste another ant belongs to and what its job is. The queen exudes an unmistakable scent. In small, young colonies with only one queen, this may characterize the smell of the nest and go ignored by the workers. Later, once the colony has grown larger, the royal eau de toilette is a clear-cut sign that the queen is at home and will not tolerate egg-laying rivals. In species where workers have functioning ovaries, this inhibits her subordinates from producing offspring of their own. If a rebellious worker does produce her own eggs, they will lack the queen's trademark scent and will be eaten by the other workers. However, if the queen goes missing, dies, or experiences a drop in her fertility, the absence of her fertile scent will ensure that the news spreads quickly through the nest, and the young workers will grab their chance. Since workers lack a sperm pouch, however, their eggs will never be fertilized, producing only sons, ultimately dooming the colony to death.

Ants use smell to identify whether a member of the same species belongs to the same colony.

It's not just Her Majesty who has a job-specific scent. Workers employed as scouts outside the nest smell different to their sisters working inside, and to the ants taking care of the larvae in the nursery. Larvae use their scent to make clear that they need feeding and caring for. Their scent is so attractive that it almost overpowers that of the colony, to the point that larvae or pupae carried off to an unfamiliar colony will not be attacked by the new colony's workers, and will be adopted instead. Major soldiers also have their own specific chemical profile that distinguishes them from minor workers. This allows those working in the nursery to tell whether the ratio of majors to minors is correct. If, for example, a large number of soldiers is killed in an attack by another marauding colony, their scent in the nest will grow weak. To balance this out, workers in the nursery will begin feeding the larvae better so that the next generation grows up to have as many large majors as possible.

The death of an ant is not immediately clear from its scent. Although a dead worker no longer participates in cleaning and feeding, she also no longer refreshes her nest scent, losing it as soon as the hydrocarbons disintegrate. At some point, a passing worker will realize that this body no longer belongs there. She will pick up her dead sister—and throw her on the garbage heap. A rotting cadaver in the nest could easily become a breeding ground for disease. Millions of years of evolution have made ants very pragmatic creatures.

THE REAL CHEMTRAILS

In the world of ants, scent serves not just as employee ID, it also forms part of an excellent chemical guidance system, which helps them to navigate the outside world.

When a worker discovers a rich source of food on her forays through the colony's territory, she is bound to be both happy and concerned. She faces several obstacles. Her size makes her unable to overpower large insect prey. If the prey is already dead, however, it is often too heavy and too cumbersome for her to manage alone. The tasty snack cannot be left lying around for long, however, or other hungry arthropods, such as centipedes, will soon discover it. There is only one thing to do: She must get help.

How she does this depends on her species and how far she is from her nest and her sisters. If help is crawling about just a few twigs and pebbles away, all she must do is emit a pheromone into the air. The other workers will follow the signal and scurry determinedly in her direction, ready to fight or provide transport. Ants that gather food—or "forage"—often make their discoveries alone. In this situation, a fleeting scent signal would be thrown to the four winds because other ants would not be able to smell which direction it was coming from, if they managed to catch a whiff of it at all. A smart ant must head off to recruit other workers. To ensure that she finds her way back to the food, she employs the same trick as Hansel and Gretel in the dark forest, or Theseus in the Minotaur's labyrinth: She leaves a trail.

Of course, when it comes to creating a trail, breadcrumbs—which immediately get eaten—do not work, nor are ants capable of spinning their own threads like spiders. Instead, they rely on a scent trail of glandular secretions. As the ant hurries along to find her sisters, she repeatedly brushes her abdomen against the ground, leaving tiny quantities of pheromones behind. And I mean, really, really tiny. We're talking about a billionth or a trillionth of a gram per yard, depending on the species of ant. To put it another way, one gram of the scent would be enough to create a trail from Earth to Jupiter, or around 20,000 times around the globe.

When our voyager finds one of her sisters at last, she uses her scent to inform her of the delicacy, drumming excitedly on her sister's head using her antennae and perhaps also providing a regurgitated sample. In some species, this is enough to recruit hundreds of gatherers in no time. Each ant also renews the scent trail as she returns, laden with food. If the source is depleted, however, she leaves no trail and the scent will soon dissipate. Soon enough, the ground will be as fresh as a blank page once again, ready for another worker to mark out a trail to a new food source.

The evolutionary origins of the scent trails used by ants to find food probably lie at the other end of the digestive system: They may have developed from the habit of marking one's territory using piles of droppings. This theory is supported by the fact that trail pheromones often originate in the rectum and rectal glands. The message "This territory is ours" may have evolved into "This food belongs to us," before finally becoming "This is the way to the food source"— not the most appetizing idea, but a very efficient process all the same.

A worker who has discovered a new food source provides her sister with a sample to help her decide whether it's worth the effort.

In addition to deliberately laid scent trails, which aid navigation, workers of most species unavoidably leave behind chemical footprints when the profile of hydrocarbons on their cuticles is transferred to the ground. These prints allow ants to identify whether they have been somewhere before, which is handy when scoping out the size of a potential new nesting site. This type of scent also reveals whether other species of ants have intruded on the colony's territory. While they defend their territory against insurgents of the same species, ants often tolerate newcomers from other species. Foragers generally avoid larger ants as a precaution, more often following the footprints of smaller species. Their trails might lead to a new food source.

Some species of ants have even specialized in reading other species' tracks. Why go to the effort of scouring the forest when you can leave the hard work to someone else? In Southeast Asia and South America, many ants of the genera *Camponotus* and *Crematogaster* have developed a partnership known as a parabiosis. Species from two groups live together as a community within the same nest, the larger *Camponotus* serving as defenders of the nest. Meanwhile, the smaller *Crematogaster* search for food and, when successful, lay down a chemical trail for their powerful friends. This kind of teamwork is rarely witnessed even among the smartest birds and mammals. Indeed, it's not always a given in human society.

DRUMS AND STRINGS

The language of scent is undoubtedly the most important way in which ants communicate. But while observing them in the laboratory or in the wild, we've noticed time and again how they touch one another with their antennae and sometimes with their front legs. They drum on one another's heads, prod each other's mouth parts, and tug at each other's mandibles. It was once thought that there was such a thing as an "antenna language." But when we record these movements on video and analyze them closely frame by frame, it's impossible to identify any pattern to the smacking and drumming. The motions and positioning of the antennae seem, by contrast, to be entirely coincidental. If this touching has any significance at all, it is perhaps only to create an increased awareness, as if the ant wants to say, "Hey, heads up! It's important!"

This method is utilized by ants from species living in small colonies of a few dozen insects when they want to lead their sisters to a food source or a new nesting site. Almost as soon as she returns from her voyage of discovery, the ant in question will use her antennae to ask another worker to follow her on what is known as a tandem run. Once the worker has displayed an interest, the ant who has made the discovery turns around and runs off. The recruited ant follows her and uses her antennae to try to touch the other ant's abdomen, primarily to maintain contact with her back legs where the ant produces an enticing pheromone via special glands. It is both fascinating and frustrating to observe this interaction because both runners frequently lose one another, and the guiding ant must wait for her colleague to find her again. Once they have finally reached their destination, the two ants either work together to harvest the food or run off again to call for backup. Eventually there are enough industrious little workers on-site with sufficient strength to carry the big lump back to the nest.

Scientists consider tandem running a form of teaching, because the ant in front guides the ant behind, showing her the way, all while giving her time to memorize the route so that, in the end, both of them know where it leads. At the end of the day, this somewhat inefficient process turns out to be worth it, even when relocating to a new nest. The tandem run takes considerably longer than if the first ant simply picked the other up and carried her to their destination, but each run increases the number of workers knowing the route, whereas an ant who has been carried has no idea which way she has come or where she has ended up. By contrast, every ant that is guided toward its destination can instruct her sisters—or help carry them—until, eventually, the entire colony has arrived at its new nest.

Besides tandem runs, ants display another important behavior, which they probably only achieve through touch: begging for food. If an ant who is hungry or tasked with the job of distributing food in the nest comes across a worker with a full crop, she need only fuss over her kindly donor with her antennae and the worker will respond by regurgitating a droplet. My doctoral supervisor, Bert Hölldobler, used a single hair to mimic the same tickling that ants create with their antennae and provoked them to regurgitate food; he was able to demonstrate that no scent signals are exchanged during this process. Simply tap an ant gently on the head and mouth and there will soon be a tasty meal in front of you. Much to the ants' disappointment, numerous parasites have also got wind of this method and have the audacity to employ this trick to sponge off well-fed workers. In a later chapter, we will look more closely at these moochers of the ant world, but we already know how they are getting their food.

But it's not all percussion—ants have an ear for strings, too. "Ear" is perhaps not quite the right term here, since ants have no ears per se, instead picking up vibrations not through the air, but through the ground. They use their legs to "hear." However, it takes great effort for a tiny insect to create vibrations in soil or wood, where ants prefer to build their nests. And then there's the fact that a signal might not even reach particularly far because the material muffles the vibrations. So how does an ant benefit from sending its message via vibrations in solids, when it could use its powerful scent-based language? Put simply, scent fails miserably at conveying the message in certain situations, while drumming and chirping can still be heard nearby.

Workers from species with small colonies lead their sisters toward a new food source or nesting site by tandem running.

Imagine for a moment that you're a leafcutter ant and your colony has been dumb enough to literally build its home on sand. Parts of the nest are constantly collapsing and one day, you're caught in the middle of it. All of a sudden, sand rains down on you, burying your little body beneath it. A landslide of this kind would crush a human being instantly, but as an ant, you have a hard exoskeleton, which copes easily with little accidents such as these. Unfortunately, the pile of sand under which you are now sitting is quite large and heavy. You will never be able to make it out on your own and the alarm pheromones, which you are using to try to call for help, can't find their way out of the sand. So, what do you do? You make some noise!

Ants have developed many different methods for sending acoustic signals. Some simply use their hindquarters to bang on the ground or against the walls of the nest, others create vibrations with their whole bodies or use their jaws to scratch at the ground. Leafcutter ants, however, are the violinists of the ant world. Instead of violins, they possess what is known as stridulatory organs. A segment of their abdomens is furnished with grooves, which the ants drag over a neighboring segment to create a high chirping sound, which is audible outside, attracting nearby workers who swiftly appear to dig their buried sister out of the sand. Saved by the bell, you might say.

This chirping emergency alarm is a special case when it comes to acoustic communication. Much like using your antennae to tap another ant, sound mainly serves to amplify a chemical message. Pheromones are simply more enticing when they are presented with a little string music; workers find it easier to follow a scent call to a food source if the ant who discovered it is also stridulating. When it comes to food, ants like to make sure they are transmitting on all available channels.

CHAPTER 5

FINELY TUNED NAVIGATION

Desert ants of the genus Cataglyphis
*are masters of navigating by celestial
compass and path integration.*

Brazil is a myrmecologist's idea of paradise. It plays host to the most wonderful species, particularly in the rainforest. On one excursion, a Brazilian student and I were in the middle of the jungle, searching cracks in branches for founding colonies of *Neoponera* ant queens when, at some point, we found we had bags full of ants but no idea which direction we had come from. The jungle looked the same whichever way we turned. "It's not as bad as it seems," I thought to myself. Perhaps some innate sense of direction would lead us the right way. All we had to do was put our heads together and decide which direction to take.

Unfortunately, another problem soon emerged: The student spoke only Portuguese and I spoke only German, English, and French. We seemed equally at odds when deciding which way to go. After some toing and froing, I came out the more persuasive of the two of us, and we set off in "my" direction. Eventually, we did manage to find our way out of the rainforest, even if we didn't end up where we had set off. It had turned out all right in the end. But in the following days, I kept coming back to the same thought: When the student and I were lost in our botanical and linguistic jungle, had she really been trying to explain to me which way to go, or had she been talking about something entirely different?

When I think back to that day, another question presents itself: If it is that easy for people to get lost, how exactly does a tiny little ant find its way home when it's too small to peer over the nearest blade of grass and can't rely on chemical trails?

GETTING AROUND THE NEST

As long as a worker carries out her work inside the nest, her world remains reasonably manageable and organized. Many ant nests contain wide corridors with narrower side corridors branching

How do you find your way in a garden, forest, or meadow when you are smaller than a blade of grass?

off toward different chambers. Since it's pitch-dark inside, eyes are of no use at all in this labyrinth of little grottoes, yet ants are perfectly capable of finding their way around. Everything they need to orient themselves is contained within their antennae. These antennae not only are able to feel how wide the corridor is, they also warn the ant of obstacles or passing ants of the same species. Ants can use their chemical senses to pick up the scent of other workers, the queen, and her offspring. They are also able to gauge temperature, the levels of moisture in the air, and the amount of carbon dioxide present. This allows the ant to determine with reasonable accuracy where she is in the nest and how far it is to the queen's chamber or the larval chamber, as each region has its own physicochemical fingerprint. It becomes moister the further underground the ant travels. The temperature remains relatively constant because the surrounding soil serves as a sort of cushion, balancing out any variations. By contrast, conditions close to the surface are more changeable because of the weather. One day it might be cool, and the soil may be wet from the rain, the next day it might be hot, the sun and wind drying out the corridors and chambers.

Thanks to its sense of gravity, the ant is also able to determine whether a corridor leads up or down. Humans and many vertebrates possess an organ responsible for this sense within their inner ear; ants, on the other hand, use their entire bodies to measure gravity. The joints between the mobile parts of their bodies boast clusters of sensory bristles. If gravity pulls her legs, antennae, head, or abdomen toward Earth's center, these limbs exert pressure on the bristles, producing a signal that tells the ant, "You are going down."

In the first days after hatching, a worker does not travel far. Initially employed as a foster mother or wet nurse, she rarely leaves the chambers housing the larvae and pupae. Though the choice of paths is limited, she ensures she does not get lost by swiftly developing a mental map of the area. In lab tests, the North American ant of the species *Formica pallidefulva* required only two to three times as long as a rat to memorize its way out of a small maze. Rats are known to have a strong sense of direction and are considered experts at navigating lab-based labyrinths.

The further the worker travels within the nest, the larger her internal map becomes. But the memory stores inside an ant's brain are scarcely enough to keep up once she steps up for duties in the big wide world outside the nest. So, she must learn a few tricks to stay on track.

A WELL-CONSTRUCTED ROAD NETWORK

Finding your way in a forest or meadow is a very different challenge from navigating a nest built specifically for ants. For starters, in the wilderness there is always the option of turning left, right, or going straight ahead. And how far left? A full right angle or just a little? How far should an ant travel before she has to decide which direction to go next? What should she pay attention to, the number of steps she has taken or the time she has been traveling? And there's another dimension to consider. An ant has no problem scaling a blade of grass or a tree, but it increases the chances of taking a wrong turn—astronomically so. A couple of signposts would be a big help. Or, better still, a well-constructed road network.

Some species, such as wood ants and leafcutter ants, have done exactly this. They travel mainly along the same paths, which they mark out with scent trails. Some of these ant roads can reach over 100 yards (90 m) in length, including little tunnels through wild undergrowth and bridges built from fallen branches or tree trunks, if a stream is blocking the way. Small troops of specialized workers keep the highways free of obstacles such as sand, leaves, and small twigs. For ants, it's less about the shortest route than the fastest route. While investigating the African Matabele ant *Megaponera analis*, Erik Frank from the University of Würzburg was able to show that they prefer circuitous routes over bare ground—over which they are able to travel twice as fast—to shorter routes made more difficult by thick vegetation.

Unlike human roads, ant highways are not fixed to the ground and may just as easily lead up over a wall or a tree trunk. If you feel like doing a little research of your own and proving that ant roads are indeed marked with scent trails, try looking in your garden, local woodland, or park and find a path well-frequented by ants on a tree. Take an eraser, rub it over the path a few times and watch how the ants react. The eraser will have removed a good amount of the pheromones, partially destroying this part of the trail, much as if the asphalt disappeared in front of us, leaving nothing but bare fields. The ants run back and forth, searching for the right scent until one of them finds it and fills in the gap using secretions from her own glands. The operation can continue as usual—and the repairs are much quicker than we're used to on our highways.

Pouring molten aluminum into a fire ant nest reveals a complex arrangement of corridors and chambers.

Ants mark out the path to a food source using a scent trail, which is used by so many workers that it becomes an established route.

These scent-based signposts have the advantage of speed as well as disappearing of their own accord once the ants lose interest in a specific route. If, for example, a herd of aphids has moved to a different branch with fresher greenery, workers will mark out and use the path to the new pasture regularly, whereas the pheromones on the old trail will fade with time. By following the stronger scent at any junction, the ants avoid taking the wrong path. Ant trails are used for as little as a few days and as long as several years, depending on the importance of the route.

BEHIND THE BEECH, RIGHT AS FAR AS THE OAK

Searching for trails in the jungle is generally futile. If you're lucky, there will still be a beaten track, but this will generally be poorly defined, quickly becoming overgrown. On my last expedition to Peru, the path through the Amazon rainforest seemed to stop abruptly at a stream. My group and I walked along the bank, but we came up against impenetrable undergrowth in both directions. We waded through the water in our rubber boots, following its course, but at no point did the path seem to continue. We discovered a fallen tree trunk, which lay over the stream; two flowering bromeliads were growing on it. Taking comfort in this lovely sight, and a few photographs to boot, we turned around and walked all the way back. We had clearly hit a dead end.

On another research operation a few days later, the scene seemed to repeat itself. Once again, the path ended abruptly and we searched for it in vain, just as before. Then I noticed a tree trunk lying on its side, two bromeliads growing out of it. It was the same sight we had seen on our previous walk, except this time we were seeing it from the other side. Now, at last, we knew where we were and where the path continued: past the bromeliads, down the stream, and then right. Nothing could have been simpler, thanks to our flowery landmark.

Ant workers do not always stick neatly to their trails, however. Some species don't care much for trails and prefer to strike out across country. So, they have to have methods for finding their way back to the nest when they decide to go off-road.

One of these methods involves observing their surroundings as they go and noticing certain landmarks, just as we oriented ourselves with the help of the tree and the bromeliads. Initially, it seems somewhat surprising that ants use these visual clues, since, as we learned in chapter 4, Communicative Sensuality, the ant's compound eyes do not provide a particularly good image, and workers of many species, such as army ants, do not have compound eyes at all. Nevertheless, wood ants can see well enough to orient themselves using striking landmarks. The African stink ant *Pachycondyla tarsata* relies mainly on sight-based navigation. If one of its workers leaves the nest, she will memorize the pattern of the treetops above her and register changes that occur as she travels on her way. Evidence proving that these ants do rely on sight (and are not secretly using their sense of smell) was gathered by Bert Hölldobler, using a clever disorientation tactic: He photographed the sky above the entrance to the nest and stuck the image above the nest, but turned the other way. Stink ants lost track of the nest as they went to work, looking at the photo and marching straight off in the wrong direction. It was as if somebody had played around with their personal signpost.

CHASING THE SUN

As capable as they are of observing photos, ants do not need to see particularly well to be able to orient themselves by sight. For many species, all they need to know is the position of the sun. The Mediterranean harvester ant, *Messor barbarus*, subsists mainly on the seeds of wild herbs, which it transports along its trails to its nest. On the journey back, it uses the sun as its compass. Depending on the time of day, it will travel at a precise angle to the sun, which will lead it back to its nest. The sun's movement across the sky, and its different position on the ant's return journey, pose no problems, though young insects newly employed outside the nest must learn to carry out the necessary calculations—without a calculator, of course.

Problems with the celestial compass do arise, however, once a myrmecologist appears with a screen, charms, and mirrors. In one classic experiment in 1911, the Swiss doctor and naturalist Felix Santschi blocked out the sun using a dark screen and reflected it at harvester ants from a different direction. The workers changed course immediately. When Santschi positioned the mirror in different locations, the ants obediently followed the simulated star. They made 180-degree U-turns without protest. These days, myrmecologists are happy to perform this test in the lab, where simple lamps take the sun's role, leading the ants the wrong way on cue. With a little finesse, they can make the insects walk in geometric patterns— with a reward at the end to make the effort worthwhile.

Wood ants, on the other hand, are not so easily fooled. They also rely on their celestial compass, and may even use the moon as a landmark, but their credulity has its limits. If the position of the reflected sun deviates too starkly from its real position, the insects notice that they are being tricked and continue marching in the original direction. Wood ants evidently have a guidance system, which triggers an alarm, overriding control of navigation. But how are they able to distinguish between the real sun and the reflected sun using their primitive compound eyes, when our complex lens eyes cannot? The answer is that compound eyes are not primitive at all. In fact, they're excellent tools for navigating by sunlight. They see something that our eyes can't: the polarization of light.

SUNSHINE ON A CLOUDY DAY

To understand how ants accomplish this, you will need to know a little more about light and its polarization. Things are about to get physical. Don't be scared: It's not that difficult. After all, ants master it intuitively.

Each light source, such as the sun or a lamp, sends out countless rays of light. A ray of light is not composed of particles like a stream of water from a garden hose; it comprises an electrical field alternating extremely fast between positive and negative. It is an electrical flicker of sorts, speeding straight through space. When a ray of light hits the eye, this electrical field energizes the electrons in the pigment retinal, at which point the molecule changes shape, triggering a signal: We see a bright flash. Our world is lit up by a constant storm of these flashes from continuous rays of light.

But let's come back to the flickering electric field of light for a moment. The positive and negative areas of the electrical field do not simply follow one another in single file; they deviate very slightly from the straight line. Imagine this like a rope: You tie one end to a fixed object and move the other up and down quickly. This creates a wave with peaks and troughs, which run along the rope—all at the same level, vertically to the floor, up and down. If you keep the rope from drooping toward the floor and continue swinging from left to right, the level of the peaks and troughs will be parallel to the ground.

Light functions in much the same way. Each peak equates to a positive area in the electrical field and each trough a negative area, and for each light beam the field oscillates at a certain level, which physicists call the "plane of polarization"—or "polarization," for short.

The polarization of light is very important when it comes to sight. In order for a light beam to stimulate a retinal molecule, the polarization plane and the elongated molecule must be suitably aligned, otherwise the light beam will simply pass through and nothing will happen. In human eyes, retinal molecules are all tangled up together and are mobile, so that every light beam can find a suitable partner. In an ant's compound eye, however, the situation is very different. The retinal molecules in some simple eyes are strictly arranged in parallel to one another. For simple eyes like these, therefore, the world only lights up if light with a suitable polarization enters the eye. These simple eyes emit a signal enabling ants to discern the polarization of light—a big advantage when you want to know where the sun is— assuming the position of the sun has anything to do with the polarization of light in the first place.

The crucial step does not take place in the sun itself, then, because the sun sends out light in all polarizations. It is much more dependent upon the atmosphere that the light beams are directed through via dispersion, falling on us from all directions. Depending on the angle from which we look at the sky, dispersion eliminates light beams more or less strongly according to their polarization. This effect is of practically no consequence when you look directly at the sun: Light beams reach the eye at all possible oscillation planes. The light is bright but it is not, ultimately, polarized. The situation changes dramatically when the sun is positioned to the side of us. If we had an ant's eyes, we would be able to tell that the light falling in this direction was somewhat weaker, but consequently more strongly polarized. If an ant has the sun at its back, the light is even weaker, and the remaining light is less polarized. Their special design allows compound eyes to act as a kind of protractor, reliably informing the ant of the position of the sun. It even works when the sky is clouded over. Only at night and during a complete eclipse, when it's really dark, will the polarization compass cease to work. On these occasions, however, ants will simply stay in the nest or stand still and wait until it is light again.

With sophisticated eyes like these, an ant could travel through the desert, where everything looks the same in every direction, and still find her way.

DESERT-DWELLING MASTERS OF NAVIGATION

In southern Arizona, the August sun beats down with a deadly heat. So far, I have been lucky enough to have only carried out excursions in mountain regions, where altitudes of 6,500 to 9,800 feet (2,000 to 3,000 m) means it never gets too hot, even in high summer. But my colleague Deborah Gordon has not been so lucky. Her key research interest is

Pogonomyrmex—harvester ants, which call flat, semidesert regions home. It is so hot there at midday that the air shimmers over the dry ground and the temperature reads over 100°F (40°C) in the shade—except there is no shade. When it comes to escaping the heat, ants and scientists are much alike: They choose to avoid the heat by heading out in the early hours of the morning. While others, like myself, are still lazing around in their huts at the research station, looking forward to breakfast, Deborah is already on her way out. She observes harvester ant scouts swarming out at sunrise and notes that those seeking food strike out as soon as they know where the decent food sources are. As the sun reaches its zenith, the ants hole up in the cool depths of their nests and the scientist retreats to the shade of her field station.

The desert ants of the genus *Cataglyphis* also spend their lives out in the heat. Where they live, it is so hot that any insect that fails to find shade or bury themselves in cooler ground will soon be fried to a crisp in the scorching sun and need only be picked up. Ant workers of this genus set out on solo trips on zig-zagging paths to forage for these tasty barbecued treats, covering an area of over 200 yards (180 m) in the deserts of North Africa. Leaving a scent trail would be pointless because pheromones evaporate instantly in the heat and, since each dead insect lies in a different spot, little is achieved by marking the route. This is why desert ants do not create trails, instead running around apparently at random. Their zig-zagging paths do in fact follow a sophisticated system: The ants prefer to travel at right angles to the direction of the wind so that they have a greater chance of catching the scent of their prey. If they happen upon a tasty morsel, they snap it up and hotfoot it out of the blazing sun into the cool of the nest. Even desert ants are likely to be fried alive if they spend too long in the sun, so the workers do not return via the winding path they set out on; they run straight back to the nest. So, how do they know which direction the nest is in if they cannot see it? After all, they're surrounded by nothing but hot, dry sand and often have no landmarks to help them get their bearings.

You've probably already guessed it: Desert ants navigate primarily by the sun—they do get more than their fair share of it, after all. But the position of the sun and the polarization of light are fairly rough guides, and the solar disk in the sky has an angular diameter of half a degree. If the ant tries to use this imprecise measure to locate its nest from 330 feet (100 m) away, it is likely that even a perfectly straight route could be out by as much as 3 feet (1 m), leading this poor little worker running past the entrance to its nest. Thankfully, desert ants have a built-in distance meter, which tells them when they should have reached it.

This meter is practically a little computer. The desert ant takes a winding, circuitous path on her way out, choosing a direct route on her way back. This means she cannot simply count the number of steps she takes and repeat that number on the way back. Instead, she must constantly check the direction she is traveling in throughout her journey and keep calculating how far her route lies from the nest, or how close to the entrance it will take her. It's possible to calculate this mathematically using vector resolution and addition, which is as likely to bring human students out in a sweat as the hot desert sun. Nobody explains to the ant how path integration, as it is known, works, but after a quick, instinctive learning curve, she will have mastered it effortlessly. Nobody knows how she does it.

Yet ants clearly know how many steps they need to take to reach the entrance to the nest. Matthias Wittlinger proved this in his doctoral thesis at the University of Ulm. He deliberately altered the mathematical makeup of some desert ants, performing experiments in which he changed the width of their stride. Working in the desert in Tunisia, he allowed the ants to find a piece of prey on a test trail in and—here is where it gets a little gruesome—would then either cut the ants legs to shorten them or lengthen them using pig bristles. The ants did not seem particularly fazed by the amputations or the stilts. They snapped up the insect and ran their precisely calculated number of steps back to the nest. However, since their legs were different lengths on their return, the ants stopped too early or too late, searching confusedly for the entrance to the nest. If, instead, Wittlinger manipulated their legs before they went searching for food, the workers applied the same scale on the way out as on the way back and found their way back to the nest without a problem. Other workers will often be waiting for the ants at the entrance to the nest, or the foragers themselves will orient themselves according to small landmarks, because even a sophisticated pedometer is not 100 percent foolproof. In a pinch, the ant will simply search around, because it knows—thanks to its handy pedometer—that the entrance to the nest must be nearby.

REWINDING THE TAPE

Along with astrocompasses and pedometers—the combination that scientists refer to as "path integration"—*Cataglyphis* desert ants also have another trick for finding their way: optic flow. By this we mean the images that flit past the eye during movement, like the landscape outside the window on a train ride. To navigate, the ant registers the film playing before its eyes as it heads out, choosing a return route that sees the same film play backward.

Ants that are carried instead of walking themselves—as is common when a colony moves to a new nest—depend entirely on optic flow. In experiments investigating how ants respond when transferred to a difference place with their eyes covered, ants are unable to find their way back to the nest.

If you attach stilts to the feet of desert ants, they take overlarge steps on their way back, missing the entrance to the nest.

This type of navigation becomes somewhat complicated if a worker has found a piece of prey that is too big to carry forward. Big chunks of prey must instead be dragged backward. When this occurs, however, the images do not appear in the part of the eye in which the ant expects them. Confused, she keeps stopping, puts the prey down, turns around and runs forward a few steps. She seems to be recalibrating her optic flow, ensuring she will know where she is going next time. She picks up her prey with new certainty and continues to drag it along, with renewed confidence.

Interestingly, the systems of path integration and optic flow seem to work independently of one another, meaning that ants have two navigational devices in their heads. This double protection can save an ant's life in the desert. Get lost and you are as good as dead.

CELESTIAL COMPASSES

Wolfgang Rössler and his team at the University of Würzburg recently discovered that at least some desert ants not only have their own celestial compass but also orient themselves to Earth's magnetic field. The researchers confused young workers of the species *Cataglyphis nova*, which were hoping to acquaint themselves with their surroundings on their first days on duty outside the nest. Ants usually take little field trips to memorize anything that might serve as a helpful landmark later when they want to find the nest's inconspicuous entrance. On these training forays, young ants spin on their own axes making little pirouettes, pausing for varying intervals. Astonishingly, during these longer pauses, they faced toward the nest, even though they were incapable of discerning the position of the sun and the polarization of the light. How did they know which way would take them home?

Rössler's team speculated that the answer could be an internal magnetic compass. Research into migratory birds shows they are guided by their own magnetic compasses, among other things, but proof of such a phenomenon in insects had been lacking until now. Scientists installed an electromagnetic coil at the entrance to an ant's nest, generating a magnetic field stronger than Earth's. When the coil was switched off, it was business as usual and the ants would stop spinning with their sights on the entrance to the nest. But when the researchers switched on the artificial magnetic field, which was constantly switching direction, or neutralizing Earth's magnetic field using an opposing external field, the young ants would start spinning and find themselves suddenly looking in different directions. The scientists were able to force the ants in the wrong direction by simulating artificial North and South Poles using a linear magnetic field. In these circumstances, the young workers went so far as to ignore the position of the sun and the polarization of the light. They were totally fixated on the magnetic field.

Desert ants prefer an orderly magnetic field, at least during the learning phase, when they spend two to three days close to the entrance of the nest. Later, when they mature into active foragers, they orient themselves predominantly by the sun. However, we do not know whether their magnetic compass is still active at this stage, or where their sense of the magnetic field is located within the body. But a few curious myrmecologists are pondering the kinds of experiments they could use to find out.

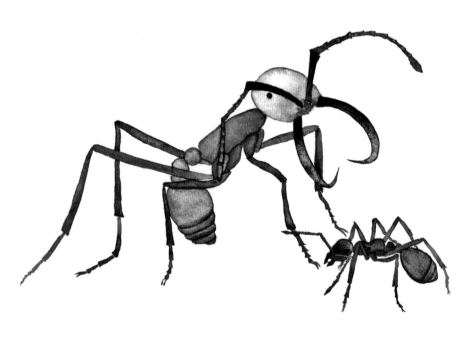

CHAPTER 6

SAVAGE HORDES

———

Army ants exhibit great variation between different worker castes. Large soldiers stand guard over the marauding swarm.

Army ants are found in the tropics and subtropics. They live all across these regions of the Americas, Africa, and Asia, where it is warm and damp throughout the year, and I have encountered them myself in the rainforests of South America and Malaysia.

Once, on a research trip, setting off for a tour of a research station near the Malaysian capital of Kuala Lumpur, my plane was delayed by twelve hours, so I didn't arrive in the morning as I had hoped, but at nine o'clock in the evening. Of course, nobody was there to pick me up and I struggled to find a taxi to take me to the station in the rainforest. I had not banked on the locals' propensity for superstition and it took a while before I managed to find a driver unafraid of driving into the jungle at night—at least, not at first. The denser the surrounding forest grew, the more his courage failed until, eventually, he turned around and began driving back. I had to use all my powers of persuasion to convince him to drop me off at the field station.

When I finally reached the facility, I was so tired I could have fallen straight into bed—if not for the news that one of my colleagues had just discovered a colony of army ants moving nests with all its belongings. My exhaustion vanished at once. Donning our headlamps, we stumbled through the pitch-black rainforest and discovered a seemingly never-ending line of ants making their way along the ground. Many of the workers were carrying golden pupae in their mandibles. When I looked more closely, I realized that the columns were not all ants; a colorful array of different creatures was traveling alongside them. A spider had joined the line, while silverfish—wingless *Apterygota*—had settled themselves on top of the ant pupae and were letting the workers carry them. There was even a little snail, transported like precious cargo. The snail species *Allopeas myrmekophilos* has developed a fragrant foam, which enables it to enjoy the luxury of a free ride. The foam causes ants to mistake the snails for their own offspring from their own nest and carefully carry them in their mandibles. The snails' pheromones are so effective that, when made to choose between a real ant larva and a snail, ants will more often pick the snail.

In two and a half hours the move was complete, my tiredness returned, and I really wanted to go to sleep, but my colleagues said I would regret missing what I was about to see. So I stayed, persevering in the damp, warm darkness. Nothing happened for some time. Then there was a flurry of activity along the ants' trail, but these weren't ants: It was a rearguard of rove beetles. They scurried along with their abdomens slightly raised, clearly following the ants' chemical trail. Rove beetles enjoy ravaging ant broods when the opportunity arises, but they did not dare to travel as part of the main caravan and risk the ants' attacking them. As for me, I was simply grateful to have arrived in time to witness a colony of army ants on the move—later that night, I finally fell into bed for some much-needed sleep.

NO MERCY

To qualify as army ants, an ant species must demonstrate certain characteristics. Unlike most normal ants, such as wood ants, who send out scouts to search for a prized food source before sending their foragers on their way, or solitary ants (such as desert ants), who take care of the entire business alone, army ants do not exhibit the same wait-and-see attitude or solitary behavior. They immediately send in powerful teams, with thousands of insects setting out on campaigns to catch prey together. Most species of army ant also live underground and go about their business in secret. This is why we know very little about them, even though occasionally we can see them hunting at our feet. The aboveground hunts of the Central and South American species *Eciton hamatum* and *Eciton burchellii* are much more spectacular.

Both species set off in much the same way. As soon as morning dawns, the ants are in the mood to hunt. They dismantle their camp and form a teeming host of bodies, beginning as a broad front but soon moving as a narrow column. On the journey, this column of hunters splits again and again: Columns of *Eciton burchellii* split into enough subcolumns to create a fan that sweeps the ground. None of the individual workers specifies which way the fan should travel. The ants are constantly running forward a little, allowing themselves to be picked back up by the mass of other ants, which continues to push forward like a tsunami. Workers transport prey to the new camp, while soldiers with threatening-looking mandibles stand guard at the edges of the column. This many-legged menace travels through the tropical forest this way at a leisurely pace of around 9 yards (8 m) per hour. If the ants come up against an obstacle, they lock together to form living bridges. *Eciton hamatum* move around exclusively via the leaves of trees, searching for the larvae of other social, state-building insects like bees, wasps, and ants. *Eciton burchelli*, on the other hand, are not picky about food; they are generalists, killing anything that falls into their clutches. This primarily includes other insects, spiders, woodlice, centipedes, millipedes, and worms, but also small vertebrates, such as lizards, toads, and young birds, who fail to escape in time. These kinds of prey would be too large for traditional hunting ants, but hordes of army ants are so great in number that they make swift work of any resistance. Their huge, sickle-shaped mandibles prevent *Eciton* army ants from cutting up large prey animals, so instead they choose to leave them lying there, much to the glee of the many flies in their entourage.

Army ants hunt in columns of thousands of insects, devouring anything that fails to get out of the way in time. These raids are overseen by soldiers with huge mandibles.

Other animals, too, are well aware that something is bound to fall off the conveyor belt when this army of ants is on the hunt. Woodpeckers and antbirds will pick out tasty treats for themselves from the motley zoo of escaping prey. Other birds of these species—and myrmecologists—rely on their calls to reveal the army swarm's location. Some antbirds have become so accustomed to this walking delivery service that they would be incapable of finding sufficient food for themselves without the help of the army ants. Army ants kill around 100,000 creatures on their raids, which can last up to twelve hours, munching through swaths of the forest until little remains.

Come evening, they set up camp—if we can really call it that . . .

A SPECIAL KIND OF TENT

A single site won't satisfy the hunger of a colony of army ants for long, so *Eciton* colonies move around constantly during this phase of nomadic wandering, which lasts two to three weeks. Many species relocate some few hundred yards away every day because their growing larvae require a constant supply of food. Given these ants' erratic lifestyle, it's not worth building a new, permanent home every time they move. Instead, they typically seek out a sufficiently sized shelter, such as a fissure in a rock, hole in the ground, hollow tree trunk, couple of tree roots, or, in an emergency, a couple of branches positioned close together. There they pitch what is known as a bivouac. They always have the necessary material at hand because a bivouac nest is composed of the ants' own bodies. Using special hooks on their feet, they join together to create a living cluster, which, in the case of *Eciton burchelli*, can be composed of half a million workers—the old insects on the outside, the young on the inside—and weigh over two pounds (1 kg), reaching over a yard (1 m) in diameter. Underground-dwelling ants such as the African genus *Dorylus* also often build a bivouac. The location of a bivouac belonging to *Dorylus molestus* makes itself known via a sinkhole in the ground, measuring over three yards in diameter, out of which the ants will have dug over 77 pounds (35 kg) of sand. Inside the bivouac, corridors branch off toward different chambers in which the queen and her offspring are sheltered. But they're not the only ones under protection.

Given the fatal efficiency with which army ants make their way through the forest, it seems almost incredible that these killers harbor such a large number of illegal tenants in their bivouacs. In the nests I have examined so far, I have found spiders, beetles, woodlice, silverfish, flies, mites, and springtails. They feed off the ants' leftovers. And that is no paltry portion, since army ants place little value on nest hygiene compared to other ants. They also do not check their bivouacs for parasites and have allowed new moochers to creep their way in over the course of evolution. It was once assumed that the different species of rove beetles, which nest alongside army ants, had the same species of origin. However, genetic comparisons have shown that different species of beetles discovered army ant nests as a potential habitat independently of one another, adapting to them accordingly.

The ants' "guests"—to speak more politely of these parasites—ability to pull this off lies in their excellent camouflage, which enables them to smell like the ants. To achieve this chemical mimicry, the parasites produce the same cocktails of scent as the ants or rub themselves against the bodies of dead ants to pick up the scent. Many parasites withdraw into quieter corners of the nest where they are less likely to draw attention to themselves. Things get tricky, however, when the hosts dismantle their camp and move on. If you can't make the journey on your own six feet, it's time to push your parasitism to the limit and let yourself be carried along by the ants transporting their larvae—just as I observed in Malaysia.

A bivouac housing thousands of ants is a practical invention in itself, but it's much more than just a mobile dormitory. If the ants are swept away by a wave during a heavy downpour, the bivouac automatically transforms into a floating raft. Their outer cuticle repels water and prevents the nest from becoming saturated. The ants' bristles also hold a thin layer of air, which gives the bivouac sufficient buoyancy to remain afloat, even amid the fiercest waves. If an ant has exhausted its supply of oxygen underwater, it frees itself from the bundle and crawls up into the open air. One of its sisters will take its place. This principle works so well that it is not only used by some army ants—such as the West African *Dorylus arcens*—but also by other species of ants such the European fire ant *Myrmica rubra* and the red fire ant *Solenopsis Invicta*, which are spreading across the northern and southern US, respectively, thanks in no small part to this strategy.

The bivouac does have one disadvantage when compared with a normal, permanent nest, however: It lacks good insulation against cold and drying environments. A thick layer of earth can absorb the heat of the day, even in summer, releasing it during colder seasons, ensuring the temperature in the soil is increasingly stable and constant the deeper underground you go. The wooden walls of a nest in a tree or a pile of twigs and needles prevent heat loss thanks to a series of trapped air pockets, which are well protected against the wind. By contrast, a bivouac is a leaky construction with thin walls, made from a material lacking in any real capacity for heat. It is essentially a draughty hut, providing only adequate protection in warmer climes. This is why army ants are only found in the hotter, tropical regions of South and Central America, the southern US, and Africa and Asia. There is only one species found in Europe, *Dorylus fulvus*, on the southeast end of the Balkan peninsula. Climate places a natural limit on the army ants' wanderlust. But, then again, the climate may very well change.

For roving army ants, it's not worth building a fixed nest. Instead, they pitch a bivouac composed of the bodies of thousands of workers.

If a colony of ants encounters a surprise flood, many species—like these fire ants—build a raft of live workers, who take turns being under water.

THE QUEEN IS DIETING AGAIN

Army ant queens represent another way in which this ferocious ant group are a little different. The queen never grows wings, never leaving her colony. When hoping to get to know a male ant, she waits patiently for her admirer in her bivouac while her daughters prepare to grant him an audience with her. The males can fly and swarm out to seek their true love, like fairytale princes. If a male is lucky, he will find the young queen's bivouac and receive the workers' approval, and be able to couple with his beloved. (Well, he and a couple of other princes, since queen army ants mate with several males.) Making their flights solo, male army ants must know how to navigate and are equipped with large compound eyes and three additional frontal ocelli. They can see very well, whereas many queens—and all workers—are practically blind. Their eyes are often just the tiny points of a single optical unit, an ommatidium, rather than part of a compound eye—and that's if they have eyes at all. Nevertheless, workers of the species *Dorylus wilverthi* are at least able to differentiate between light and dark, probably via light sensors in their skin.

There is another reason why *Dorylus wilverthi* is a particularly interesting species of ant: It holds two world records. The first is for the species with the largest colonies, numbering over twenty million workers. The second is held by its queens, which are the largest among all ant species, measuring over 2 inches (5.2 cm)—about the length of a small mouse, without its tail.

The queens of the *Eciton* species are not quite as large, but they still tower over their workers. They are largest during the stationary phase, when the army ants have not moved on for two or three weeks, choosing instead to stay in one place. Their insatiable larvae will have now pupated and are no longer in constant need of food, so the hunters do not need to go out as frequently and the surrounding territory is sufficient to feed the colony for a time. The queen uses this period of calm to produce new eggs. During the nomadic phase, she was slim and agile, but now she is practically unrecognizable. Her abdomen swells to gargantuan proportions until she is too plump to cover long distances unaided. Instead, she lays thousands of eggs in a great feat of strength, producing the next generation of workers. She has to hurry, because as soon as the young workers have hatched the fresh larvae will have to be fed and she will have to snap back to her traveling weight. Then the next nomadic phase will begin. Army ants are not great at putting their feet up.

THIS LAND IS MY LAND

Occasionally, there will be a couple of special insects among the many offspring that hatch during the stationary phase. Even the queen will die someday, and since each colony only has a single queen, the colony itself will meet its end shortly afterward. The workers will occasionally fatten up a few of their sisters into young queens, allowing them to continue to grow until one is chosen to establish a new colony. This is why potential candidates do not strike out on their own in the hope of establishing their own colony or being accepted by another colony. The lifestyle of a pillaging horde requires a great army of workers and soldiers from the very beginning. The chosen one will receive a good portion of the present royal household as a dowry and the colony will split—but only in two. There are, however, several young queens, so they do not all benefit from a colony of their own.

How the ants decide which young queen takes the crown is reminiscent of the process on a reality TV show. Each candidate has her supporters among the workers, either growing or shrinking in number. The candidate with the most votes receives half the kingdom and trots off together with her entourage, hoping an eligible male will soon arrive on the scene. Her spurned sisters, on the other hand, are no longer wanted in the colony and are isolated. They and any loyal supporters stop receiving food and are condemned to perish miserably—the tragic high point of any good reality TV show.

Both halves of the colony, now two colonies, go their own ways. If, one day, they encounter each other or another unfamiliar colony of army ants, they will not come to blows despite their aggressive attitudes. Instead, both parties will turn and walk off in different directions. Since they always travel on foot—even their queens can't fly—army ants cannot surmount larger natural obstacles like rivers. Even a narrow forest clearing just several hundred yards wide will keep them in check if no food is available to them on the way. The same applies to a mountain or a range of hills where it is too cold for them. This is why army ants' habitats often form little islands of their own within larger, barren areas. They're also generally not found on ocean islands or islands in large lakes, since their queens do not fly and cannot be carried there on the wind. They might get around a lot, but army ants typically choose to stick to their home turf.

SAUSAGE FLIES AND BITING SUTURES

Not all species of army ants have the same lifestyle, though. We know almost nothing about how the underground species live: Only some species have ever been seen by a handful of scientists. Ants of the South African genus *Aenictogiton* are so secretive that nobody has seen one of its workers or queens to this day; only males of this genus have appeared from time to time. Even so, at least the scientists recognized them as ants. The great naturalist Carolus Linnaeus—whom we have to thank for the two-term naming system, initially believed the males of the ant species *Dorylus wilverthi* to be wasps, and there are probably still some males that have been miscategorized as their own species because they bear so little resemblance to the females in their colonies. In some areas, *Dorylus wilverthi* are said to resemble sausages on the deli counter, hence their common name: the sausage fly.

Army ants don't always find themselves on the menu, however. For one thing, they are incredibly useful for tending to gaping wounds. Old advanced civilizations presumably knew about this practice, which is still used by the Maasai today: The ants' mandibles are perfect for stapling together tears in the skin. An ant is held up with its mouth facing the wound, and once it bites down, it is decapitated.

Army ants aren't just fascinating, they're useful too—so long as you're a human and not an insect destined for their dinner table.

Oh, and then there's the matter of *Jaglavak* . . .

THE PRINCE OF INSECTS

From our perspective, the lives of the Mofu people of Northern Cameroon may seem hard and spartan. The villages of this small tribe lie just under 220 miles (350 km) south of Lake Chad on the border with Nigeria, on a mountain so dry that few wild animals live there, save for insects and a few birds. It is here where the Mofu grow sorghum, one of the few crops capable of tolerating such harsh conditions. They store their harvest in roundhouses with walls made of wooden beams, large stones, and air-dried mud, all covered with straw roofs. They are constructed in much the same way as the houses in which the Mofu themselves live. Despite—or perhaps because of—the simplicity of their lives, the Mofu lead a contented existence. In fact, sometimes their only concerns seem to be that the rains will not come or that termites will invade the village.

Like ants, termites are eusocial insects, living in huge colonies with a single queen laying endless numbers of eggs while workers and soldiers take care of everything else. The area where the Mofu have settled is home to a species of termites

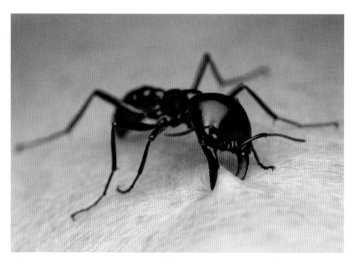

The army ant's powerful mandibles are an excellent defensive tool, but they are also used in traditional medicine to suture wounds.

that feed on wood and seeds. In the termites' eyes, the Mofu villages are akin to paradise, but when they descend on these gardens of Eden, the villages affected have hell to pay.

The largest part of the termites' nest is located underground but they tunnel into the walls of the Mofu's houses, hollowing them out until the infestation is so severe that sections of wall crumble away at the slightest touch. If the termites reach the roof, they will nibble away at the straw until the entire structure caves in. By far the worst thing, however, is when the termites get stuck in last year's sorghum stores. They are capable of destroying not just the villagers' food but also the seeds for the coming season. It's no wonder that a visit from termites can mean absolute ruin for a village. When it comes to fighting back, there's only one solution: Jaglavak, the driver ant.

The village elders know to proceed carefully with Jaglavak, the so-called Prince of Insects. After all, ants are not particularly pleasant colleagues and have a nasty bite. First, a colony must be found—no easy undertaking, since this species of ant is one of the underground-dwelling varieties of driver ants, making its nests hard to trace. It's an ideal job for curious children. Sent on their way by their elders, the children fan out to search for the ants. If they are successful, they call for help from the adults, who use hooks to break up the nest and place some of the colony into a sealable calabash. Back in the village, the villagers click their fingers to welcome Jaglavak as an honored guest. The village elders put down ocher at one of the entrances to the termites' nest as a gift to Jaglavak and ask him to eat only the termites, not the inhabitants of the house or their goats. Once that's done, all the villagers can do is wait.

Meanwhile, drama is unfolding in the tunnels inside the termite hill. The driver ants use their sense of smell to determine whose kingdom they are in and plunge at once into the fight of their lives. Using their mandibles, they seize the considerably larger soldier termites by the legs and antennae, incapacitating them. Other ants then tear through the immobilized soldiers' armor. In no time at all, the ants break through the outer defenses and make their way toward the royal chambers. The workers inside the chambers, caring for the parents of the colony, know that danger is close at hand and try to carry the parents to safety. They push and pull, finally succeeding in moving the royal couple to the main chamber. Soldier termites press their heads together to seal off the corridors behind the royal couple, but the ants are small enough to slip through the gaps. The battle enters its next stage and often continues for several days.

The Mofu are not aware of any of this. They are not remotely interested in what is going on beneath the ground. For them, all that matters is that two or three weeks after asking Jaglavak for help, the termites have completely disappeared, and the ants have gone with them. The village is saved. All hail the Prince of Insects!

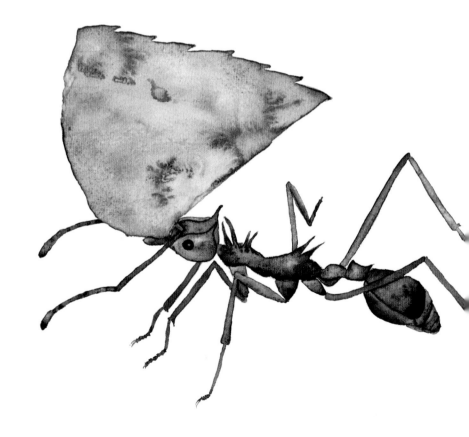

CHAPTER 7

A GARDEN FOR A CITY OF MILLIONS

*A leafcutter ant carries
a piece of leaf.*

One morning, while I was working on my PhD, I encountered something strange. I was the first to arrive at the institute that morning and immediately noticed a procession of red ants marching along the corridor in a long line. The workers were all carrying small, semicircular pieces of rubber mat in their mandibles, holding them aloft as they marched from the genetics laboratory to the climatic chamber where their nests were located. Clearly, our leafcutter ants had escaped in the night and got right to work gathering new material for the colony. Since we had no potted plants to speak of, they had no choice but to find a substitute, something at least faintly reminiscent of leaves. I couldn't imagine what they might have chosen. It wasn't until I closed the door to the laboratory and followed the trail of industrious little workers that I saw a group of their sisters, busily cutting the aforementioned pieces; they had scaled the gel dryer and were busy carving the rubber mat loose from its cover, producing neat, manageable little pieces. They were letting the pieces drop to the floor where they were picked up by other workers and carried away. It was the same behavior we see leafcutters exhibiting in nature, with real leaves. But I had to interrupt their work as soon as possible, and set about capturing them and returning them to the climatic chamber. Thankfully, the damage they had caused was insubstantial. But from that day on, and for several years afterward, we had to patch up the holes in the gel dryer cover whenever we came to use it. In the evenings, we would double- and triple-check to make sure that the nest chambers containing the ants were firmly secure.

THE FLOWER CHILDREN OF THE ANT COMMUNITY

When leafcutter ants are not busy dismantling laboratories, they can be found in the tropics and subtropics of the Americas. The most northerly of the forty-seven species that we know of can be found in the warm states of Louisiana and Texas, with the most southerly living in Patagonia in Argentina. They do not exist in the wild in Europe, Africa, or Asia, but plenty of zoos are keen to host these entertaining little critters and keep colonies, which astound wide-eyed visitors of all ages with their antics as they march along transparent plastic pipes holding their leaf-flags aloft. Even branches of large outdoor suppliers use leafcutters to draw in customers, confident they will appreciate a glimpse of these inspiring insects when they come in to buy sleeping bags and raincoats. There's no doubt about it: Alongside army ants, leafcutter ants are the pop stars of the ant world.

The biting soldiers of both taxonomic groups have been used by primitive peoples to suture open wounds, and can also be found on the menu in some regions. In Mexico and Colombia, *Hormigas Culonas* is the name given to a dish of fried leafcutter queens, a delicacy and aphrodisiac—and a somewhat unusual honor for a species of insect that only has sex a single day in its life. So what is it like? Crispy, salty, a little nutty and slightly earthy. A doctoral student once brought along a sample from South America of red and black ants without their heads, legs, or antennae. They weren't to my taste, but perhaps that was because, on some level, I felt sad these young leafcutter queens would never have the opportunity to establish a new colony.

Aside from these few similarities, army ants and leafcutter ants are polar opposites in the rich repertoire of ant species. While army ants are meat-eaters without a fixed abode, leafcutters subsist on a vegan diet and live in huge, permanent nests. And what nests they are! But sometimes even a million-ant metropolis can start to feel a little cramped.

BIRTH OF A MEGACITY

Young leafcutter queens establish their colonies in the traditional way. They couple with at least five males on their nuptial flights, filling their sperm pouches with around 300 million sperm. They then break off their wings and seek out a suitable site for the new colony on foot. They prefer open country with little vegetation where they can easily dig, such as can be found on the edge of country roads. When she arrives, the mother-to-be buries herself a foot under the ground, excavating a small chamber. On her last trip outside, to leave a mouthful of soil outside the entrance, she takes one final glimpse of daylight. But she has no time to savor the moment, she has plenty of work to do inside the chamber. Not only does she need to spend the coming days and weeks laying the eggs, which will become the first generation of workers, she also has another job to do, one which is at least as important for the colony's fortunes: She has to do some gardening.

In a special pouch inside her mouth, the infrabuccal pocket, the queen carries a little fungal mycelium as a souvenir from her old nest. This isn't any old fungus; it is a special species on which all the leafcutters' work depends, as we are about to see. If the fungus dies, the queen might as well give up, because leafcutter ants cannot live without it. Without this fungus the queen and all her workers would soon starve. She places the fungal threads on the ground, fertilizing them with a little liquid feces—an unseemly but very efficient method, proven over millions of years to boost growth. Only then does she retreat to lay her first eggs.

In the coming days, the queen will alternate between tending to the fungus and her offspring. The fungus soon develops into a thick carpet, but both the queen—living off her now useless flight muscles and her fat reserves—and the larvae, who are fed eggs instead—are forbidden to touch it. Only the workers will be allowed to snack on the fungus, after they've hatched. In return, however, they will be put to work in the garden. They will dig their way out via the sealed entrance and begin to gather leaves, which they cut up and carry back to the chamber for the fungus to use as a substrate. Only two or three in every hundred queens make it to this point, after which they pass on the running of the day-to-day business to their daughters, enabling queens to concentrate fully on egg-laying.

The queen's lonely hermitage is little more than a small family home for now, but the aim is to create a giant state that will function as a superorganism.

A THOUSAND ROOMS AND THREE MILLION INHABITANTS

In its first year, the new colony is still struggling for survival. In its second year, it begins to gain momentum, growing more steadily—and not just in terms of the number of ants, but in variation of the workers' body sizes as well. Whereas the first generations were still fairly puny, subsequent larvae gradually develop into more imposing ants, until eventually there are four subcastes of workers: First, there are the small "minors" who spend their whole lives in the nest, where they tend to the fungus, scuttling about between its threads thanks to their diminutive size. Then there's the somewhat larger minors and "medias," the two most numerous of the subcastes, who take on practically all routine jobs from gathering leaves to defending against hostile ants. And finally the "majors," or soldiers, into whose rearing and care the colony funnels so many resources that only colonies with over 100,000 insects can afford these giant, large-headed ants. Despite the exorbitant costs, the investment is worth it because majors can cut through leather using their mandibles and can also bite through human skin without issue, as countless myrmecologists can confirm. I must admit that humans are at fault in these situations. Leafcutter ants are fundamentally peaceful creatures. But it's best not to start digging up their nests.

In an established colony, all the workers live together in one nest whose residents are so vast in number, it would be like burying Penn Station underground, along with the population of Baltimore, three or four times over. Scientists like the Brazilian researcher Luiz Forti have taken drastic measures to discover what a metropolis such as this might look like, filling whole nests with cement and then digging up the resulting sculpture. It sounds simple enough, but in terms of research this was a major feat. The first challenge was convincing a building company to deliver 6.9 tons (6.3 metric tons) of cement and 440 gallons (2,000 L) of water to the middle of nowhere. This then had to be stirred up to create a mixture thin enough to flow through the narrow corridors inside the nest without blocking them. Once the hill was full to the brim, they had to wait for the cement to set three whole weeks. Then came the delicate work of digging it out with a shovel. A digger would have been too imprecise and destroyed part of the nest, and then all their efforts would have been in vain. So, like archeologists, the researchers were obliged to dig out this once thriving civilization by hand. After several gallons of sweat had been spent, however, it became increasingly clear that the effort had been worth it, so much so that the researchers found themselves digging out not just one nest, but several.

This cement cast gave us a reasonably precise image of the construction of a leafcutter nest. The central portion occupied a surface measuring around 55 square yards (50 m²), equivalent to around eight parking spots. This doesn't seem like much, but ants apply the same principle as New York does when squeezing its residents into the tranquil district of Manhattan: They build vertically. Their corridors and chambers reach down as much as 26 feet (8 m) underground. A nest generally comprises 1,000 to 2,000 chambers, but the record stands at 7,864 cavern-like spaces, ranging in volume from that of a tennis ball to that of a soccer ball. Researchers found

Leafcutter nests are vast in size. The workers build them without a plan but are sure to include a sophisticated ventilation system and efficient waste disposal.

ants or fungus gardens in a third of the chambers; the others served as garbage dumps.

But the rooms comprise just one part of this fascinating phenomenon; the infrastructure, composed of different kinds of special tunnels, is just as astounding. The corridors leading to the chambers are built with a gentle incline, to keep out water during a heavy downpour. But this alone is not enough to keep the insects and the fungus alive, since both species take in oxygen and give out carbon dioxide through respiration. Without a sophisticated ventilation system, these underground inhabitants would suffocate. This is why the nest itself is surrounded by a circle of steep channels leading up in an arc to the surface from the deepest point, ending in chimneys allowing for plentiful ventilation. Between them run several miles of underground corridors, and some of these tunnels lead to exits and satellite chambers as much as 110 yards (100 m) away. If we include these outposts, large nests can cover an area larger than several soccer fields.

The whole structure is so complex and sophisticated that you would be inclined to suspect there were skilled architects involved, but ants build all of this without a plan and without complex computer simulation. They rely on their little brains, the delicate sensors on their antennae, and a few dozen million years of evolution.

Between two and three million ants live in these megacities. An estimated eight million ants of the leafcutter species *Atta sexdens* has even been recorded in one city. And they all depend on one very special fungus.

THE FUNGAL ISSUE

Leafcutter ants have the same problem as many vegetarians: They cannot digest cellulose. Unfortunately, however, plants are largely composed of precisely this, so vegetarians would have to go hungry if they couldn't delegate the work to someone else. Rabbits, cows, and, to some extent, humans all harbor microorganisms in our intestines, which break down cellulose into other substances that we can then absorb. Alternatively, we simply digest the microbes themselves whole. To manage this with their short intestines, leafcutter ants have joined forces with a mushroom species from the genus *Leucoagaricus*. They regularly supply their fungus with fresh plant substrate, keeping it moist and clean. In return, the ants harvest the nutritious swellings on the ends of the fungal threads, known as gongylidia. The larvae are completely dependent on this protein-rich food, whereas workers can make do with leaf sap in a pinch.

That's the deal, anyway, but the ants must work hard to ensure it is implemented. First, they send out their scouts to find suitable shrubs and trees in the surrounding area. They pass over plants that are toxic to the fungus, instead seeking out juicy leaves or sometimes colorful flowers. How the scouts know what the fungus needs at a given time remains a mystery. They have practically nothing to do with caring for the fungus, after all, and so do not see which plants the fungus responds to and how. They are nevertheless able to select the ideal fertilizers with uncanny certainty, creating a scent trail running from the plant back to the nest.

The gang of workers is then able to find the site and begins a cascade of finely tuned tasks seen nowhere else in the animal kingdom. The leafcutters begin proceedings. They climb the plant due for harvesting and slice off large pieces of the leaves using their powerful jaws. Their technique is closer to that of a Japanese saw than it is to a pair of scissors. One side braces itself against the leaf, like a free hand against a cutting board, while the other slices upward through the greenery, creating a half circle. If the worker deems a leaf to be particularly suitable, she will attract her sisters to it by making a knocking sound with her abdomen. Leafcutter workers often let the cut pieces simply drop to the floor, where transport workers will be waiting for them.

What follows is the popular and well-known relay race back to the nest, with the ants carrying their leaf flags. The average transport worker holds aloft a piece of leaf that, in human terms, would weigh around 660 pounds (300 kg). She takes a nice, neat route back to the nest, at speeds worthy of a world-class marathon runner. If the road is long, she will pass her relay flag to a colleague, who will cover the next portion of the journey, until the last runner finally drops the leaf at the entrance to the nest. At this point, the minors pounce on the leaf and tear it into smaller pieces, which are then gradually shredded. The minors chew the shreds up into a pulp, which they feed to the fungus threads or hyphae.

Leafcutter ants take their name from their habit of cutting leaves into pieces.

According to scientists, the leaf-harvesting process can number as many as twenty-nine steps. It's a tremendously laborious act when we consider that, although some workers take in a little sap during this activity, leafcutter ants do not ultimately live off the leaves; they merely use them to feed their fungus garden. The work and welfare of this whole million-insect colony depends on this fungus. It's no wonder that there is an army of tiny specialists tending to the gardens day and night. And, funnily enough, it's the smallest subcaste that shoulders the responsibility for the colony's survival.

ADVANCED GARDENING

It is no accident that minors are so tiny. They have to be. They are the colony's gardeners and must be able to creep into any secret corner of the fungal labyrinth, the mycelium. This uses its interwoven filaments to form a system of caves, reminiscent of the pattern of holes in a natural sponge. The minors crawl along the corridors, checking the condition of the fungus.

They feed it with chewed-up leaves and lay the groundwork for new cultivation areas. These new areas are fertilized with liquid feces. Every now and then, the gardener will nibble the bulbous protein-rich gongylidia at the tips of the fungal filaments and distribute it to the larvae and her sisters. It's a matter of give and take, in which both sides stand to gain.

But even the best gardeners can't work in peace if nasty weeds have other plans. A fungus grower's sworn enemy is the sac fungus *Escovopsis*, which lies in wait, seeking an opportunity to infest the culture and bring about the ants' ruin. Even in young leafcutter colonies, one in fifteen fungal cultures will be infected, and the parasite will spread rapidly until, within a year or two, half of the fungal cultures will be affected. In particularly severe cases, the infestation may be so bad that the colony will have to leave its nest—the project into which it has poured many years of graft and effort—and start all over again.

To keep matters from reaching this stage, minors carry out patrols throughout the network, removing any spores or fibers from foreign fungi. Workers on business outside the nest, who

Transport workers carry fragments of leaf to the nest.

could be carrying all kinds of pathogens on their cuticles, are barred from entering the fungus garden. And dead ants, the remains of leaves, and dead mycelium find their way into special waste disposal chambers. The work of refuse collection is taken up by old workers, who turn the waste material frequently so that it breaks down more quickly. Much like in human societies, this important job is largely thankless: Waste disposal workers are avoided by their sisters and are not allowed to move freely within the nest, to ensure that they do not spread *Escovopsis* spores. But these preventive measures are not enough to keep the parasites away from the fungal monoculture. For this, the ants have another trick up their sleeve—and a special ally.

Like any true chemistry expert, leafcutter ants regularly turn their chemical talents to their gardens. Minors of the species *Acromyrmex octospinosus* produce a secretion, composed of over twenty components in the metapleural glands located at the back of their thoraxes, which they then spread over the fungus. The substances secreted include the growth hormone indoleacetic acid, which stimulates the fungus to proliferate as quickly and as densely as possible. The secretion also contains antibiotics and antifungals, which the ants cannot synthesize themselves. The real manufacturers are symbiotic bacteria, which, depending on species, live on different parts of the ants' bodies and feed off glandular secretions. Thus, leafcutter gardens often witness wrestling matches among four different participants unlike any other in the animal kingdom: The ants tend to a fungus, which is attacked by another fungus, which the insects fight with the help of bacteria. And this underground battle has raged for fifty million years. Millions of years ago, ants cultivated the first fungi, setting in motion a development that was so successful that neither the leafcutters nor the fungus are able to exist without it. And both the parasitic fungus and the bacteria have been part of this setup since almost the

Leafcutter ants do not feed on leaves, instead feeding off a fungus, which grows on chewed-up plant material. Only small minor workers can fit through the tiny hollows between the fungus threads.

beginning; one side is constantly thinking up new tricks, which the other side counters. It is an example of a coevolutionary arms race with an unusually high number of participants.

IT'S WHAT'S ON THE INSIDE THAT COUNTS

Leafcutter ants owe their underground gardening successes to commitment, chemistry, and a raft of allies. But what about their farming habits above ground?

These can be summed up in two words: Pretty lousy! Leaf-cutters will happily defoliate whole fields of citrus and other fruit trees, cacao, cotton, coconut, and many other crops. A colony takes a mere twenty-four hours to strip a lemon tree, consuming as much greenery as a cow. The ants cause 130 million dollars' worth of damage per year in the Brazilian state of São Paulo alone. And there is almost no way of stopping them. The only thing that helps is a protective ring of waste from the waste disposal chamber in the ants' nest. The fear of carrying parasitic fungus back to their own nest will keep the ants at bay for a good month or so. But since there is no option of ordering this wonder cure online, anyone hoping to use it must first dig through to the ants' waste disposal chamber.

It's not just plantation owners who have a low opinion of leafcutter ants; cattle breeders are not keen either. In some parts of South America, some leafcutters have selected grass stalks as their substrate of choice for feeding their fungus. These ants have sharp spines on their backs to keep from being accidentally swallowed by grazing cattle as they harvest the grass. Biting into one of these little wandering cacti causes pain in the lips and tongue, so cattle avoid pastures where leafcutters are present.

So, it's for good reason that leafcutters are not popular with farmers. Yet, despite their uncompromising harvest practices, they have an important role to play in nature. Like all ants living in underground nests, the corridors they build help to aerate the ground and by carrying leaves and associated fungi in and around the nest, they add countless nutrients to the soil, which would otherwise have only been available on the surface. This process of fertilization makes the jungle floor up to ten times more fertile. Ants are highly valuable to the ecosystem, provided you stay out of their way.

SMALL BUT DEADLY

Farmers are by no means the only enemies leafcutters have to contend with. They are considered a quick and easy snack by anteaters, armadillos, lizards, and birds. Thanks to their flag-waving processions, leafcutters are easy to spot and their nests are easy to find. Animals with powerful claws, like anteaters, can easily rip through the top part of an ant nest, located above ground, otherwise these hungry predators would have to make do with the ants that trot past them on the road, as if carried along a conveyor belt. Leafcutters are not especially keen on either of these options, so they send their fighters to the front, to take on the attacker. Soldiers head into combat against huge foes such as anteaters and armadillos, armed with their powerful mandibles. These are capable of biting through thick skin causing, at the very least, a good deal of discomfort. The leafcutters meet smaller attackers, such as unfamiliar ant colonies, with an army much greater in number, composed of the minor subcaste. Conflicts between colonies of the same species can involve thousands of fighters and last several days. It's the size of the army that makes all the difference.

When it comes to waging war on another group of ants, however, leafcutters are mostly obliged to hang back. Ants of the genus *Azteca* live in close communion with certain plants, feeding from them and defending them against pests in return (see chapter 8, A Treehouse for the Nation). Since *Azteca* ants are smaller than leafcutters and smaller in number, they have developed another strategy to defend their convoys: They build pitfalls. To do this, they gnaw canals into branches, biting a hole that leads to the surface, and lie in wait concealed inside these tunnels. A simple branch can be as holey as Swiss cheese—and as dangerous as a minefield. If a leafcutter steps into one of the holes, an *Azteca* will bite the ant's foot, gripping it firmly. As the leafcutter attempts to wriggle free, it stumbles into the other holes. Soon it can no longer move, making it easy prey for the *Azteca*. Scouts rarely return to their colonies after a visit to an *Azteca* tree. The pitfall method works so well that even army ants give such trees a wide berth and large insects such as grasshoppers have no chance of escaping.

A particularly intimate threat is posed to leafcutters working outside the nest—a threat from the air. Scuttle flies of the genus *Pseudacteon* love to lay their eggs on the thoraxes of leafcutters, who are unable to defend themselves while carrying their little leaf flags. Once hatched, the fly larvae crawl inside the ant's head, where they begin eating their host from the inside out. In two weeks or so, little is left of the brain,

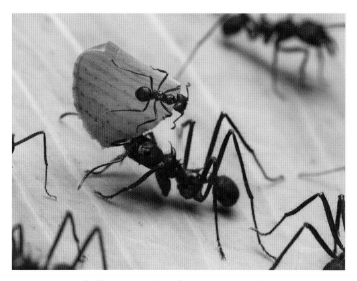

Parasitic scuttle flies love to lay their eggs on workers who cannot defend themselves because they are busy carrying leaves. Minor workers sometimes ride along on the leaves or the carriers themselves to provide protection and fight off the flies.

such that the ant worker can no longer orient itself, rambling aimlessly every which way. A few weeks later, the ant's head falls off entirely, which is why scuttle flies are also known as ant-decapitating flies. The larvae pupate in the detached heads and new flies emerge soon after.

It does not have to come to this. To nip this sinister process in the bud, transport ants are often accompanied by air defense workers. Tiny minors ride along on their larger sisters, or on the pieces of leaf, and defend them against attacks from scuttle flies. Between these airborne assaults, they cleanse the leaves of harmful microorganisms. The minors themselves have a natural form of defense against the decapitating flies: They are simply too small to serve as suitable bassinets for the flies' larvae.

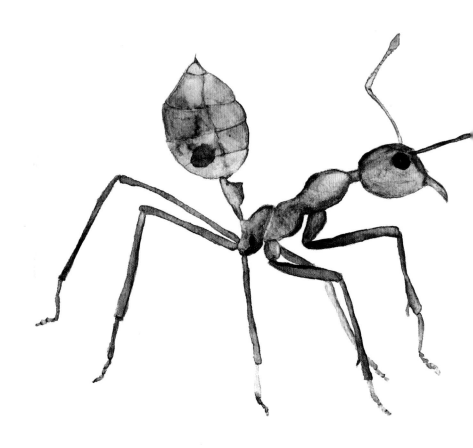

CHAPTER 8

A TREEHOUSE FOR THE NATION

Weaver ants of the species Oecophylla smaragdina *are also known as green ants, due to their green-colored abdomens.*

The gardening habits of ants do not stop at fungi. When it comes to lavishing care on certain plants, these fastidious little creatures are some of the most successful gardeners in the animal kingdom—except for humans, of course. They don't even shy away from clearing large patches of woodland or creating monocultures. And when it comes to using natural fertilizers, they are experts.

Essentially, it's quite a simple arrangement: The plants provide the ants with a place to live, and often feed them, too, and in return the ants nibble away at anything that might threaten their host or compete with it for resources. Ant plants, or myrmecophytes, are plants that cooperate in a mutually beneficial way or, in scientific terms, engage in mutual symbiosis with ants. And it seems to be an attractive offer, with over one hundred species of plants from a variety of families and several hundreds of species of ants developing these close relationships.

THEFT = FOOD

The partnership between the pitcher plant *Nepenthes bicalcarata* and its ants is one of the more baffling symbioses, which at first glance might seem to suggest that the ants exploit their host without offering it anything in return.

Pitcher plants face a common problem in that they grow on surfaces that offer them little in the way of nitrogen. Without nitrogen, plants are likely to suffer from atrophy, since this element is an important component of basic molecules such as proteins, DNA, and chlorophyll. Since beggars can't be choosers, pitcher plants have found a somewhat unusual solution: They get their nitrogen through dishonest means by catching and eating insects. The insects will have feasted on other plants, taking in their nitrogen and incorporating it into their own bodies. This practice makes insects nitrogen-gatherers in a sense—and for carnivorous plants, that makes them little walking nitrogen supplements. Hoping to benefit from these valuable dietary supplements, pitcher plants have transformed some of their leaves into the pitcher-like shapes from which they take their name: cup-shaped with extremely slippery sides, filled with digestive fluid. The pitchers bear sweet nectar around the opening, attracting insects that slip in as they try to get a taste, falling to their deaths. It's a downright diabolical construction that works exceptionally well and provides the pitcher plant with its extra portion of nitrogen—provided nobody else tries to steal its prey.

But this is exactly what happens to *Nepenthes bicalcarata*, a pitcher plant endemic to Borneo. Its stems are semi-hollow and it is in these hollows where ants of the species *Camponotus schmitzi* delight in building their nests. These ants have adapted completely to living alongside their hosts. They have adhesive pads on their feet, known as arolia, which enable them to walk over the pitcher's smooth surfaces without slipping, and their bodies are covered in long-chain alcohols, which act as a lubricant, making it easier for them to get in and out of water. The ants use this ability of theirs to commit reckless acts of burglary. They crawl into the pitchers and pull nice, fat chunks of prey out of the plant's digestive juices. They let themselves sink into the liquid, spending up to thirty seconds diving for tasty treats, and without coming to any harm. The ants take several hours to heave their bounty the few inches to the rim of the pitcher, as they struggle to find purchase. Eventually, however, they take their quarry to the nest and the pitcher plant is left with nothing. Or is it?

As a matter of fact, on closer observation, pitcher plants are not the pitiful victims they appear to be. On the contrary, they help the ants to rob them. The leaves of this particular species of pitcher plant are not as slippery as other pitcher plants. This is the only reason that ants are able to climb in and out without slipping. Their digestive juices are also less lethal than those of other species. They are not as acidic and contain a lower concentration of enzymes, such that even some frogs are happy to try a quick dip. And if that weren't enough, the pitcher plant also provides its ants with nectar, producing extra solely for its guests. In doing so, it reduces the effectiveness of its deathtrap and repays the ants for stealing its prey. So, what does it get out of it?

The species Camponotus schmitzi *survives where others die: in flesh-eating pitcher plants. They have even been known to steal the plants' prey.*

The answer lies in the balance of coexistence with and without the ants. While the ants are entirely dependent on the pitcher plant, the pitcher plant itself can survive without them. Without them, however, it does not thrive as well. This is due to a few easy overlooked processes, where the ants intervene, which benefits the plant. For example, sometimes larger prey animals succeed in freeing themselves from the pitcher at the last moment, escaping along with all their precious nitrogen. The ants stop them from fleeing by snapping them up. They also catch the larvae of flies and gnats, which might otherwise continue to grow inside the pitcher, taking in nitrogen from the digestive juices and using it to grow to maturity before ultimately flying away. Instead of disappearing, the nitrogen taken from the plant ends up inside the ants, meaning that it is not quite lost after all. These special ants are real homebodies and prefer to spend their lives hidden away in their nests in hollow stems. This is where they bring their prey, where they store their waste, carry out their business, and where they die. The end products of all these processes contain nitrogen compounds, which are released as they decompose and are absorbed by the plant via its cells inside the stems. Ultimately, the ants do not steal any nitrogen from the pitcher plant; they simply transport it out of the pitcher, where it's not safe, into the stems, where it is. The plant takes in between half to three quarters of its nitrogen in this indirect way. And the ants also provide protection against plant-eating weevils and remove fungal spores and other pathogens. If that's not a good deal, I don't know what is.

WILL GARDEN FOR ROOM AND BOARD

The relationship between ants and their plants is seldom so refined, however, although other species of plants appreciate the nitrogen that ants bring. The trumpet tree *Cecropia peltata* takes almost all the nitrogen it requires from the waste produced by ants of the genus *Azteca*. Like the overwhelming

Plants, which live in symbiosis with ants, often offer their guests hollow spaces, known as domatia, as nesting chambers.

majority of ant trees, which very quickly fill empty gaps in the jungle in tropical areas of South America, the trumpet tree also keeps up its side of the bargain as host to the ants, primarily to protect itself against pests and predators. The *Azteca* often successfully fend off attacks from leafcutter ants, as we saw in the introduction, Small but Mighty. In return, the trees offer these societies of insects a safe place to live inside their hollow trunks, which we scientists call a domatium, or myrmecodomatium, which means little more than "ant-dwelling" but sounds much smarter. Since the ants could always find pleasant accommodation elsewhere, the ant trees raise their offer to include little balls of food on the undersides of their leaf stalks. These are known as Müllerian bodies and are rich in protein and fat, the ideal fortifying food for ant larvae. Mature ants mainly satisfy their need for nutrients with the sweet excretions of soft scales, which the ants keep for milking. Only in colonies without farming of soft scales do ants nibble on the bodies themselves.

Aside from hungry leafcutters, the *Azteca* provide protection against another danger the trees face. Epiphytes, or air plants, have worked out a strategy for managing the competition for light, allowing them to grow as high and as quickly as possible: They simply rent a penthouse suite, colonizing the upper branches of trees. Although epiphytes, unlike parasitic plants like mistletoe, do not extract nutrients from their host tree or steal its sap, they prove a burden on ant trees. They overshadow the trees' own leaves, preventing them from taking in enough light to photosynthesize, and their weight can also cause branches to break. If the tree is home to an *Azteca* colony, however, it has no need to worry: The ants snap up young epiphyte seedlings as mercilessly as they do vines and creepers.

Despite all the benefits, this exchange does bring a disadvantage for the ant trees: Its little guests attract woodpeckers, which hammer away at the trunk and branches to get at the tasty ants and their brood. Still, this arrangement seems to work overall. Although both parties are able to survive alone, they prefer to do so together.

If in doubt, it helps to make others depend on you.

AN ARMY OF ADDICTS

Some ants take gardening very seriously. *Pseudomyrmex ferruginea*, for example, view anything that approaches its charge as a threat—and will attack at a moment's notice. All it takes to trigger the alarm is the scent of an unsuspecting myrmecologist, carried on the wind as far as the ants' home, the bullhorn acacia. Soldiers patrol along its branches day and night and scouts run around the base of the tree. In a heartbeat, thousands of workers come streaming out of its big, hollow thorns, where the colony lives, descending on researchers, goats, or cicadas. In return for the ants' protection, the tree offers living chambers inside its stipular spines, which reach up to 7 inches (17 cm) in length and look like buffalo horns; it also provides highly coveted balls of food (named Beltian bodies after Thomas Belt, the man who discovered them), as well as nourishing nectar located at the tips of its leaflets. So far, so normal, we might think. But that is where we'd be dead wrong. It's the nectar, above all else, that allows the acacia to keep its guests coming back.

It might appear that the acacia tends to the ants with great care. The workers cannot digest sucrose, which we know as plain old household sugar, and amazingly the sweetness the ants take in from the acacia is, unlike the nectar of other plants, free from sucrose. However, our excitement at this act of neighborliness only lasted until researchers associated with my colleague Martin Heil brought these ants into the lab and new workers hatched out. It then emerged that the ants were not intolerant to sucrose on hatching; in fact, they coped well with sugar. At least until they got a taste of the acacia nectar. From that moment on, their digestive systems refused to process the sucrose. Suddenly, the enzyme responsible for digesting sugar, invertase, became inactive in the laboratory ants. It had been sabotaged, as Martin Heil and his colleagues discovered. And the culprit? It was the acacia tree! This seemingly attentive friend to the ants has a particularly perfidious and effective method for making itself indispensable: The tree secretly mixes a second enzyme, chitinase, into its nectar, putting the ants' invertase permanently out of action. Once an ant eats it, it becomes instantly addicted and can no longer tolerate any other nectar, leaving it dependent on acacia nectar for the rest of its life. But why does the plant do it?

As with many relationships, it has a lot to do with jealousy. Now, the acacia does not know what it's doing to the ants, of course, and it certainly hasn't made its plans for sabotage in cold blood. But as the ants and acacia have coevolved, it has obviously paid off for the tree to bind its protectors to it by force. Acacias are snacked on with much greater ferocity by other insects, and overrun by faster growing plants, when they

The inhabitants of the bullhorn acacia take the job of guarding their host plant very seriously. All it takes is one researcher to put the colony on the alert.

are unable to attract the ants to them. They are then obliged to invest lots of energy and resources in producing effector molecules. An ant colony is a much more cost-effective and aggressive solution, because the ants can only survive as long as the tree does. They fight with their backs against the wall every time and plunge into battle whenever needed. And yet, only very few creatures can survive on the acacia. They are home to perhaps the only mostly vegetarian spider on the planet.

Bagheera kiplingi takes its scientific name from the black panther in Rudyard Kipling's *The Jungle Book*. Much like its namesake, this little jumping spider is pretty smart, dwelling on old branches of the acacia where there is little nectar or fruit, and making it of little interest to the ants. Very occasionally, a patrol might mistakenly find its way into the spider's domain; otherwise the *Bagheera* is free to enjoy the peace and quiet. The spider only ventures onto the younger branches when it needs to feed. It does whatever it can to avoid the ants, even taking the precaution of moving like an ant. Its goal is to reach the Beltian bodies, which make up the majority of its food. On occasion, it will also sip on the tree's nectar, and from time to time it may also eat a couple of ants or their larvae, and little flies. Even an almost-vegetarian has its moments of weakness.

Meanwhile, the ants of the acacia do not just keep the tree free from clinging plants, they also ensure that no other little plants come near it. They munch away all shoots within half a yard of the trunk. This little sphere of influence gives you a rough idea of how rigorously ants deal with unwanted vegetation elsewhere.

INSIDE THE DEVIL'S GARDEN

Chullachaqui is not much to look at. He's short, with one leg shorter than the other and a foot that points backward. He doesn't mind, of course; according to stories told by the Quechua, Chullachaqui can take the form of another person or animal whenever he likes. Posing as a family member, he lures people into the jungle or leads hunters down the wrong track, disguised as their prey. Sometimes he will even kidnap a child, raising it to adulthood, until it becomes a Chullachaqui, too—a ghost or a demon that lives in secret clearings in the forest and punishes anyone who breaks its laws.

But no myrmecologist worth her salt is likely to let that stop her.

My colleague Deborah Gordon is as unfazed by legends as she is by the heat of the Arizona desert, where we met her earlier in chapter 5, Finely Tuned Navigation. She devoted four years of her life as a researcher to the question of how the labyrinthine devil's gardens of Peru and Brazil came to be, which are said to be the work of Chullachaqui. In the heart of the jungle, surrounded by trees, shrubs, creepers, and brush of all kinds, they appear before you all of a sudden: bright clearings, often home to a single plant species and nothing else. This is often the tree *Duroia hirsuta*, from the coffee family, but other gardens boast plants from the melastome family of the genera *Tococa* and *Clidemia*. The locals do not know how these monocultures have come to be; all they know is that no human is responsible. In their legends, they have made Chullachaqui the creator of these gardens. The truth? It's all the work of one species of ant.

The lemon ant *Myrmelachista schumanni* has a contract with its host plants: The thickets of branches and stems of *Duroia*, *Tococa*, and *Clidemia* offer it sheltered domatia for its nests, while in return the plants receive a private army prepared to defend them against all kinds of hungry herbivores. They do tend to get a little carried away, however. When a lemon ant encounters a shoot belonging to an unwanted plant on its forays, it transforms into a little six-legged demon. It bites a hole in the plant tissue and fills it with formic acid. This tiny quantity would usually cause little harm, but the ant is not alone, it has as many as three million sisters—and thousands of units of formic acid make an exceptionally effective herbicide, as Deborah was able to prove. Within hours, the tissue along the plant's main vessel dies off, and the shoot will be dead within a week. This fate awaits all plants that try to gain a foothold in a devil's garden. The ants do not want them there, however beautiful or rare they might be. These devilish little gardeners will not even tolerate unfamiliar plants with domatia suitable for *Myrmelachista* nests. As far as they are concerned, their host plant must reign supreme. Nothing else is allowed. And so, their gardens become monocultures, comprising several hundred *Duroia* trees and covering an area the size of two basketball courts. It is monocultural monotony in the midst of tropical diversity.

Of course, a garden of this size is a project sustained over many generations, but this poses no problem for *Myrmelachista schumanni*. Its colonies can boast as many as 15,000 queens, each producing new offspring—such that these societies can

It's not just humans who clear the rainforest and replace it with monoculture. These monotonous devil's gardens in the heart of the jungle are the work of the Myrmelachista schumanni ant, which build their nests in swellings in plants, known as domatia (smaller image).

essentially live on forever. Going by its size, Deborah was able to calculate that the largest garden she examined was just over 800 years old. It's certainly a successful arrangement that the ants and the plants have come to. Or is it? After all, if you do a deal with the devil, you have to pay the price!

Despite being created by ants, devil's gardens suffer the same curse as man-made monocultures: They make a tempting buffet for hungry insects. The treetops within the gardens are less dense than they would be in a wild jungle, making these plantations brighter, almost spotlighted, inevitably drawing the attention of passing insects. When, for instance, a cicada is drawn in by the light, it is greeted by dozens of the same food plant, crammed into a small space. This is paradise for any insect that lives off *Duroia* leaves, of course; they have no inkling of the devil's role in any of this. The ants make every effort to defend their plants, but even a colony numbering a million insects cannot be everywhere at once. And so, the *Duroia* trees inside the gardens are actually more likely to be nibbled at than solitary trees elsewhere in the rain forest. Deborah speculates that this may be what has prevented the ants from turning the entire jungle into one giant devil's garden over time.

Yet this coexistence with the ants must be of value to the trees, which produce extra hollow swellings for the ants, to serve as dwellings. A tree would never tolerate the ants if they were doing more harm than good, as we have seen in the ants' treetop gardens.

A FLOWERING TRASH PILE

Many ants that live in harmony with plants put their heart and soul into biting predators and clearing weeds. But what makes a weed a weed? In the ant world, as in the human one, it's a matter of opinion. Some species develop partnerships with epiphytes, which spread along the higher branches of trees and shrubs. Their ant gardens are not just raised beds; they're whole little cities in the clouds and are sometimes home to several ant species at once.

The enterprise often begins with the seed of an epiphyte, such as a bromeliad, fig, or *Philodendron*, left behind on a branch by a bird. The bird's digestive tract has little effect on the seed, which will germinate within a few days and cling to the substrate beneath it using aerial roots. From then on, it will be fairly secure, but it will need water and nutrients because its roots do not reach the ground below. If the little plant is lucky, it need not wait too long to be rescued. Its savior will appear in the form of an ant, such as *Camponotus femoratus*, to whom the plants' network of aerial roots offers the perfect structure for a cozy nest. The ants quickly haul soil up to the plant, cramming it between the root fibers, or use a kind of papier-mâché made of chewed-up wood and plant remnants to build a cardboard nest. The roots provide the paper-based dwelling with stability and secure it to the branch. Once it is secure, it can grow from modest proportions to the size of a football. The soil that the ants have fetched also offers other residents a guaranteed minimum level of moisture. Fungi permeate the ant garden with their fibers and add to the stability of the construction. Eventually, a regular storm will not be able to shake the structure. Since individual species of ants prefer different locations for their gardens, the canopies of some forests bear a diverse range of hanging, architectural designs, often decorated with pretty flowers, just like window boxes.

Some epiphytes reward the ants for their aerial gardening exploits with nutritious natural produce. The ants feed mainly on the sweet secretions of aphids, which they keep not far from their nests. But they are not likely to scorn the extra nectar, which the epiphytes produce especially for them. Some air plants also furnish their seeds with little appendages known as elaiosomes. If an ant encounters one of these seeds, she will carry it off to her nest, where the workers will consume the nutrient-rich elaiosome. They are not particularly interested in the seed itself, which will be disregarded and will soon germinate. Some species of ant fertilize and water the seedlings in their nest, such that they grow large and eventually break through the nest wall. In doing so, they perpetuate the expansion of the garden, because the ants soon fill the gaps between the new branches and roots with soil. Thus, an ant garden is a colorful process of give and take among different animals, plants, and fungi, with everyone contributing to the community and everybody getting something out of it. At least *almost* everybody. The tree that serves as the basis for the ant garden comes away empty-handed, but at least the treetop-dwelling ants don't go so far as to misappropriate their leaves. Other species are less fussy in this respect—and a little more green-fingered.

Some species of ant create floating gardens in the trees and care for them avidly.

A CITY IN THE TREES

Not all ants are a drab yellow, reddish brown, or black. There are red ones, like Jaglavak, responsible for liberating the Cameroonian villages from termites in chapter 6, Savage Hordes, and there are green ants, too. Yes, green. The queens of the species *Oecophylla smaragdina* have bright green abdomens like fresh peas. Their workers aren't quite as fashionable in this respect, but they are still green enough to have earned the name "green ant" in East Asia and Australia.

Oecophylla smaragdina is one of only two species of weaver ant alive today. Strangely, the other species—*Oecophylla longinoda*, found in the African tropics—is not green at all. In fact, it's somewhere between yellow-brown and light red. Between 23 to 56 million years ago, the weaver ant family was far greater in number, with some of the ants spreading throughout Europe where they would occasionally become trapped in droplets of tree resin, which over time became amber and preserved the ants for our era. Unfortunately, however, it's not known if these extinct species really were the great weavers their name suggests. Despite the many drops of amber in existence today, there has never been a piece big enough to contain whole colonies.

Still, today's weaver ants do not hold back when it comes to their art. They can't afford not to weave because their colonies are extremely large, sometimes comprising over half a million workers. All those workers need housing, but since weaver ants live in the upper canopies of trees—where sufficiently sized hollows are few and far between—they have hit upon an amazing DIY method for making homes of their own.

When it's time to move to a new nest because the old one is damaged or too small, or if the ants simply feel like a change, the large workers of the major caste swarm out, looking for a suitable spot for their new home. But they don't go crawling into dark corners or dingy hollows. They explore the local tree canopy, tugging at the leaves with all their might. When one major finds a leaf that bends, nearby majors join in and get to work. Their aim? To create an artificial hollow using several leaves. They have to pull the edges of the leaves together, which is no easy task. At first, the leaves are often so far apart that a single ant can't reach both edges at once. So, without further ado, weaver ants will use their bodies to form a chain from one side to the other. One ant bites onto the edge of a leaf, the other grips hold of its petiole (waist), a third seizes hold of the second, and so on until the last ant in the chain can use the claws on its legs to cling onto the edge of the other leaf. And then they pull together as one. A single chain is often not enough, so more workers create parallel chains. Eventually, even the strongest leaf will yield and curve far enough that the ants are able to hold both edges side by side. The worst bit is over.

Then comes the "weaving"—though "gluing" might be closer to the truth. The builder ants can't use their mandibles to hold the leaves together forever, so other workers hurry off to one of their existing nests and return clutching a few larvae in their final larval stage. During this stage of their development—and only this stage—the larvae have silk glands that produce fine but durable threads. The larvae usually use this silk to spin their cocoons, enabling them to transform from caterpillar-like creatures into adult ants in peace. But weaver ants have discovered a better use for it. The worker tasked with "weaving" holds its larva directly over the edge of the leaf and drums on its head with its antennae, causing it to begin producing silk. The larva responds with utmost calm, as if it knows precision is the name of the game. The worker moves the larva to the other edge of the leaf, until a silken thread spans the two edges. The worker swings back and forth, drumming a few more times on the larva's head to let it know to produce more silk. The larva itself is like a shuttle on an old loom, carrying the weft from one side to the other.

It can take a few hours to a day before all the leaves are glued together, forming a comfortable hollow. The ants furnish the inside of the hollow with internal walls made of silk, and then the new nest is ready to go. The exhausted larvae, now with no silk left to spin their own cocoons, are granted a place of honor inside the nest, where they pupate naked, in a sense. But their colony will not abandon them now that they have sacrificed themselves so selflessly: The protection offered by the nest of leaves is more than adequate for successful metamorphosis. What it lacks, however, are minor workers, who are generally responsible for tending to the brood. These ants practically never leave the nest on their own initiative; at most, they will

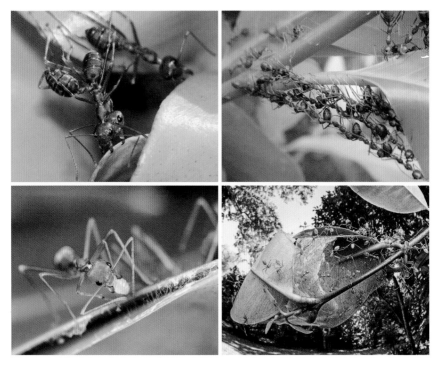

For weaver ants, building a nest is a team effort. First, a suitable leaf is bent into shape. Then the ants seize hold of one another to create a chain of bodies, pulling the edges of the leaf together. Other workers are tasked with sticking the edges of the leaf together using silk produced by ant larvae. The finished nest is just one of several dozen inhabited by one colony.

venture a couple of steps out the door to milk the aphids kept by the colony. When the colony moves, the minors will be picked up by a few majors and carried to their new home.

The newly built nest is not the colony's only dwelling. Each mature colony has dozens or even hundreds of nests, which can measure up to half a yard (50 cm) and are distributed over several trees. The outer limits of the colony's territory are inhabited primarily by older workers, who remain in their "barracks," watching to ensure that no foreign colonies of weavers stumble onto their patch, because if there's anything that weaver ants hate, it's neighbors from the same species. Each ant uses little brown droplets from its anus to carefully mark out its territory. There are also demilitarized zones between territories, which no worker of either side is permitted to cross. Nevertheless, encounters do happen from time to time. Rivals threaten one another with their mandibles open wide and their rear ends steeply raised, and—when they must—they fight.

The people of Southeast Asia have been using the weaver ants' aggression for their own ends since at least 304 CE. The oldest records can be traced back to this time and were kept by fruit growers who distributed ants across their plantations to exterminate pests. These snappy little helpers are still relied upon to keep trees clear of pests, so much so that farmers are happy to forgive them for biting them during the harvest—especially since the ants bring them a little extra income. In some regions in Asia, ant queen larvae are considered a delicacy, costing as much as twice the price of a good cut of beef. The worker brood is not quite so valuable and is sold as high-quality bird feed, known as *kroto*. And of course, ants are also much-coveted ingredients in traditional Indian and Chinese medicine, where they are used as an aphrodisiac—as well as to treat rheumatism. In Australia, with a little luck, you can even buy a caipirinha mixed with weaver ants instead of lemons. Life among humans isn't always easy.

CHAPTER 9

MILKING IT: ANTS AND THEIR LIVESTOCK

An ant milks an aphid to harvest its sweet honeydew.

The greatest revolution in human history did not take place in France, nor did it have anything to do with steam engines, electricity, or even the internet. No, the most important radical change that humanity ever witnessed was so fundamental that it barely occurs to us anymore. It was the shift that saw us go from being nomadic hunter-gatherers to settled farmers and livestock breeders. Where our ancestors once had only been able to eat when they had been lucky on the hunt, they were now able to rely on steady provisions thanks to their livestock. And they were able to fill far more bellies as a result. Over time, small settlements, villages, and towns developed, where people lived off—and with—their sheep, goats, cows, and pigs as guarantors of their prosperity and riches. For humans, this Neolithic revolution took place between 10,000 and 15,000 years ago, depending on the region. But ants discovered livestock rearing and agriculture several millions of years before any of us.

MANNA FROM HEAVEN

A settled community must know how to source sufficient food. Hunting alone is not enough for a colony with many mouths to feed. A large area of territory will be stripped bare in no time at all. What this army of sisters needs is a cornucopia of sorts, providing an endless food supply. Or, to put it another way: In many cases, only farming can support the existence of a modern ant colony.

The ants' livestock of choice has six legs and sits on a plant, feeding off its sap. In scientific terms, these creatures are known as Hemiptera, which sounds perhaps not nicer, but at least a little more refined than their more rustic common name, "true bugs." We humans are never keen to encounter Hemiptera in our gardens because they include aphids, cicadae, and other bugs we consider pests; in the ants' eyes, however, they are the most valued creatures under the sun. And, in a very literal sense, they are the sweetest, too, thanks to their sweet honeydew, which ants harvest.

This sweet sap originally comes from plants that aphids and other creatures tap using their probosces. The channels inside leaves and stalks are pressured, which squeezes the liquid inside into the insect's mouth so that there's almost no need to suck. It sates the insect's need for energy-rich sugars in a flash because the sap itself comes directly from leaves, which are actively photosynthesizing, enabling the plant to transform sunlight, water, and carbon dioxide into sugars. As with sugary sodas, the concentrations of essential amino acids and proteins in these saps is extremely low. Thus, these little sap-suckers must drink huge quantities of sap to secure a more balanced diet, filtering out the more valuable nutrients. Aphids take in more than their own weight in sap—per hour! And those that drink a lot have to find some way to get rid of it, too.

To keep from exploding, aphids release the excess as honeydew from their bottoms. Some species simply hurl it as far away from themselves as possible, so as not to find themselves smothered by the sticky substance. Occasionally, these sticky droplets can be found smeared on the windscreen and paintwork of cars unlucky enough to have parked beneath a heavily infested tree. Or they dry out overnight, forming little white granules that happen to be prized as a delicacy in the Middle East. In the Book of Exodus, this substance is said to have saved the lives of the people of Israel. It was the original manna from heaven, which gave them sustenance on their forty-year odyssey through the desert. Europeans, too, love to spread honeydew on their toast. It arrives on breakfast tables

The aphid's favorite snack is the amino acids in plant sap. They excrete the excess sugar water, providing ants with welcome nutrients.

under names such as forest honey or honeydew honey, but not until it's been through an aphid's digestive tract and refined in a bee's honey stomach. It may have been eaten twice already, but it still tastes good to us.

It's hardly surprising that, at some point, many ants realized that honeydew is a wonderful energy boost. Perhaps they stumbled across a couple of droplets at first, drawing them closer to the aphids until symbiosis developed, an arrangement that benefits them both. Meanwhile, ants are often careful to ensure there are always adequate numbers of "their" plant-suckers and that other, unfamiliar beneficiaries do not hassle them. Any creature that gets too close to the herd—for instance, to lay its eggs inside the aphids, as some ichneumon flies are known to do—will be bitten and repelled with full force.

And that's not all ants do for their six-legged honey cows.

ON LAND AND ON WATER

There's a pond in my garden. It's nothing fancy but it's a pretty little spot with frogs, irises, and a few water lilies; water lilies with ants running around on top of them. There is no way the ants have built their homes there; the nest must be somewhere in the garden, on dry land, and yet every day a few workers can be seen crawling along over to the floating leaves and white flowers. Now, ants are not known to be great swimmers. They do not go straight under, because their cuticles are water-repellent and the hairs on their bodies create a thin layer of air, which acts as a custom-fit lifejacket. And yet, in the absence of effective technology, it's an enormous effort for these insects to travel in a specific direction in the water. Nevertheless, they take the risk of jumping from an overhanging willow branch onto a water lily, continuing their journey toward the floating pads.

Every now and then, one will slip and land in the pond and take great pains to escape the water. The ants in my garden are clearly not going through all this effort just for the fun of it. There must be some specific reward, something drawing them in, something that makes all the effort they go through day after day worthwhile. But what is it?

The answer is small and black and can be found sucking eagerly on the leaves and flowers of the water lily. Aphids sometimes develop wings, enabling them to reach new host plants by air. There must have been at least one pregnant female cast up on my water lilies one day and somehow, the ants got wind of this new, difficult-to-reach herd. The ants decided to farm the water lily colony—as if there weren't enough aphids elsewhere in my garden. A few workers will clamber industriously from lily pad to lily pad and "milk" the aphids in their floating pastures by drumming their antennae on the aphids' bottoms and taking the honeydew the aphids secrete. Talk about a well-earned drink!

TEAM PLAYERS

This daily journey across the pond allows us to witness just how hell-bent ants are when it comes to accessing the fruits of their livestock, and it also demonstrates the extent to which these sap-suckers can rely on the ants' care, even under the most difficult of circumstances. Many species of aphid count on their shepherds to such an extent that they have even adapted their anatomy to live in symbiosis with ants. They invest less in defending themselves than aphids that live without ants. If you've got ants protecting you, you can afford for the wax film coating your body to be thinner and can make do with two little toxin-producing horns on your bottom as your active defense mechanism when things get desperate. Aphids have even adapted their honeydew secretions to this life of symbiosis. Instead of simply hurling a droplet somewhere where it will not be picked up, the aphids hold on to it using bristles above the anus, until an ant removes it. If, contrary to expectations, no ants make an appearance, the aphid draws the droplet back in and saves it until the next milkmaid appears.

The ultimate example of Hemiptera adapting to their ant-shepherds can be found among the mealybugs of the genus *Hippeococcus*, which are native to the Indonesian island of Java. They have developed long, grasping legs with feet resembling suction cups. When the bug is sitting undisturbed on a stem, it's not altogether clear why it needs these legs. But if it feels threatened, it will quickly climb onto the back of the nearest ant and cling on with its claws. Mealybugs that are too small to do so, or who have missed the signal to flee, are picked up by the ants in their jaws and hurried out of the danger zone. When the honeydew is at stake, ants are prepared to serve as trusty steeds.

ADOPTED BY ANTS

The adjustments ants make in their roles as shepherds are less anatomical in nature, but some species have rearranged their whole lives to care for their livestock. Some even live almost exclusively on honeydew, abandoning hunting altogether. From time to time, they treat themselves to a couple of aphids when the herd is big enough or when the colony is in urgent need of protein. The decision as to whether to indulge in the milked honeydew or an aphid steak depends on the time of year and on the species of ant. This gives the aphids another highly effective incentive to keep their shepherds in honeydew—so they don't go getting any ideas.

The queens of several species of the genus *Acropyga* take a pregnant mealy bug with them as a dowry on their nuptial flights. This ensures they have something sweet on hand as soon as they begin establishing their new colony, and means that their young brood is provided for. Ants often keep livestock in their own nests when necessary. In winter, when the sap-suckers are at risk of freezing, the ants bring them into the warm. During colder periods, the cornfield ant *Lasius neoniger* cares for the eggs of its corn root aphids as well as the eggs of its own queen, and if the colony has to move, it transports the aphid young just as carefully as you would fresh eggs. As soon as the weather warms in spring, it sets the new generation of aphids down directly on the juiciest seedlings, shoots, and roots.

But even ants in the tropics, where it is warm all year round, work intensively to rear their livestock. It's for this reason that mealybugs of the species *Malaicoccus khooi* in Malaysia are no longer remotely capable of rearing their offspring alone. They rely on the support of their ants *Dolichoderus cuspidatus*, in whose bivouac nests they lay their eggs and where the brood grows in the care of its foster family. The ants, numbering around 100,000, have practically adopted their 5,000 mealybugs. They spend the whole day carrying these demanding little suckers from A to B, always on the lookout for the freshest pastures. The nest and the feeding site can be as much as 22 yards (20 m) apart. And if one location's plant sap is no longer good enough, the entire nest uproots and moves on to new pastures. *Dolichoderus* and several related species have shown that ants invented the practice of nomadic shepherding long before humans even existed.

ALL GOOD THINGS COME TO AN END

So, do the ants and their sap-sucking friends enjoy this state of bliss forever? Occasionally, but sometimes you only spot the bars on your gilded cage when you try stretching your legs.

The black garden ant *Lasius niger*, found crawling across the gardens and parks of Europe and some of North America, is a devoted defender of the black bean aphid *Aphis fabae*. It provides a comprehensive support program, including defense against predators such as ladybugs, green lacewings, and hoverfly larvae, as well as removal of sticky honeydew. You would think the bean aphid would be grateful for this, but all is not quite well with this seemingly happy couple: A closer look reveals that the aphids are much more active when the ants are not around, suddenly becoming too lazy to move when their shepherdess appears.

A team of British researchers wanted to know whether this was a fluke or whether there was more to it. They devised an experiment to investigate this using clearly defined parameters and prevented their bean aphids from walking over the usual leaves, instead making them travel over fine laboratory filter paper. And lo and behold, the test aphids traveled across the fresh paper at a speed of 0.5 millimeters per second; by contrast, under the supervision of the ants, they moved considerably more slowly, covering 0.35 millimeters in the same time. And what's more, there was no need for a single ant to be present to create this slowing effect. All it took was for an ant to have been sitting on the filter paper beforehand. All it takes is a whiff of their masters to keep the black bean aphids from running away.

The results of this experiment fit well with the other measures that ants take to keep their herds together: They emit chemicals that suppress the growth of the aphids' wings; if an aphid nevertheless succeeds in growing them, the ants simply bite them off. Ants love their livestock, but above all they love that they are *their* livestock—and they do all they can to make sure it stays that way.

But woe betide them when the aphids decide to fight back!

THE OPPRESSED BECOMES THE OPPRESSOR

It's nothing personal. It's not a matter of revenge and it's not calculated. In nature, sometimes a creature will get hungry and nibble on the things around her, learning what is and is not edible. If a creature discovers a new food source, she has a slight edge over the competition when it comes to successfully releasing offspring into the world. This leads to behaviors at which we humans can only marvel, and sometimes shake our heads. The aphid *Paracletus cimiciformis* has developed one such behavior.

In its home in the Mediterranean, the *Paracletus* perches on the branches of the pistachio tree, but in my native Germany it has been forced to make do with grasses. It is here where it allows itself to be cared for by the pavement ant *Tetramorium caespitum*. This symbiosis becomes increasingly complicated as the aphid progresses through its life cycle, taking no less than nine different forms. As well as numerous winged and wingless varieties, and varieties that reproduce asexually or by mating, there are two types of larvae that are genetically identical but activate different genes as they develop, which means they ultimately look different. The little green larva laps up sap from its host plant like a good pet and deposits honeydew. Its twin, by contrast, has a brown, flattened body and is a specialist when it comes to surviving the winter. The aphid larva produces a pheromone that enables it to smell exactly like one of the ants' own larvae. In its efforts to carry its lost offspring to safety, the ant promptly carries the aphid into the nest and tucks it into the nesting chamber alongside its unhatched siblings. And thus, this little Trojan horse achieves its goal.

As soon as the aphid larva's belly starts rumbling, it casts about for some nearby sap and, since there is no grass in the nesting chamber, it uses its proboscis to tap the ants' larvae and drink their hemolymph. It does not kill its victim, but this bloodletting might not be all that beneficial. The ants will be glad when spring arrives and the offspring—both ant and aphid larvae—can be transported back to the warmer upper sections of the nest. Here, the aphid larvae will have access to the grass roots, a more species-appropriate source of nutrition. They will transform into chubby little adult aphids and live the rest of their lives as vegetarians.

But aphids are not the only animals that live more or less in harmony with ants, nor are they the most brutal.

KEEPING CATERPILLARS

Life is no picnic for a butterfly caterpillar. It must eat as much as possible in a short time without ending up in the stomach of one of its many predators. To make it to adulthood at all, caterpillars have developed an impressive variety of creative defense strategies: They use their abdomens to imitate scary faces, cover themselves in a thick cloak of long hair, or are so poisonous that any schmuck that makes the mistake of swallowing one will never forget how sick it made him. Or they hire ants as their personal security service.

Three quarters of all species of gossamer-winged butterflies use ants as their personal bodyguards. When the time comes for the female to lay her eggs, she seeks out a suitable food plant growing close to a suitable ant colony. The Idas blue *Plebejus idas*, for example, finds a gorse bush located in the territory of the black garden ant *Lasius niger*, while the chalkhill butterfly *Lysandra coridon* prefers horseshoe vetch and the pavement ant *Tetramorium caespitum*, or the red wood ant *Formica rufa*. Butterflies are generally choosier when it comes to plants; the species of ant is less crucial in their view.

Once the little caterpillars have hatched out and been discovered by the ants, they are safe. Many species of ant even build them little stables out of soft earth, into which they carry their nocturnal livestock to prevent them from attracting predators during the day. And it works. In experiments where researchers have prevented ants from protecting the caterpillars, less than a quarter of the gossamer-winged butterflies generally survived, and often they were all eaten. When it comes to bodyguards, you can rely on ants—provided they are appropriately remunerated for their services.

It's not just aphids who enjoy partnerships with ants. Gossamer-winged butterfly caterpillars also receive their protection in exchange for sweet honeydew.

And this is where it gets difficult for the caterpillars. Unlike aphids, they do not feed on the sweet sap of their food plants, instead eating the plant tissue. This means that they do not automatically produce sugar water, which the ants can then accept as payment. Instead, the caterpillars must knuckle down and synthesize honeydew especially for their ants, using their Newcomer's gland developed especially for this purpose. They then secrete this through a gap in their backs. This makes no difference for the ants, but it does for the caterpillars: producing honeydew takes a lot of energy and material, which they then lack for their own development. The caterpillars stay smaller than they would be without the ants and even as adult butterflies, which have developed under protection, they are not as large as their peers who develop without the ants. Personal security does not come cheap, even in the animal kingdom—and it can be exploited, too.

BEHIND ENEMY LINES

In each ant colony, one place is safer than all the others. The nesting chamber is the Fort Knox of the insect world. Insects that succeed in being carried into this space and not eaten can sit back and relax in the knowledge that they're safe. We've already seen that some aphids overwinter in their shepherds' nesting chambers, and there are also some gossamer-winged butterflies—like the Alcon blue *Maculinea alcon*—which use scent to disguise themselves as ant larvae, allowing themselves to be carried around and even be better fed than the ants' real offspring. This isn't enough for the Large Blue *Phengaris arion*, however. When it starts feeling hungry in spring, the time has come for this wolf to hatch out of its sheep's clothing: this coddled and cared-for house pet secretly tears into the ants' offspring, feasting on the larvae. This begs the question of who is really serving as livestock in this situation. After all, the ants do not benefit from their interactions with the gossamer-winged butterflies; the butterflies themselves make the ants work for them and, to top it off, they also eat the ants' offspring. In the struggle for survival, gratitude does not figure into it.

Whether a caterpillar has behaved well in the nesting chamber or not, at the end of their time as caterpillars, all gossamer-winged butterflies must face the same problem: The techniques they employed to trick the ants into believing they, too, were ants only work while they are in caterpillar form—once they become adult butterflies, they're viewed simply as a welcome source of ant food. The butterflies get a brief reprieve while undergoing their metamorphosis in the heart of enemy territory as pupae, which are of little interest to the ants. What follows, however, rivals a spy thriller in terms of drama, in which the unmasked agent tries to make it back to her own lair. Early in the morning, when the ants have yet to become particularly active, the gossamer-winged butterfly hatches out of her cocoon and tiptoes out of the ants' nest and into the open. If she remains unnoticed, she will soon be able to dance over the meadows, mate, and commit a new generation to the care of the ants. If she is spotted, however, she will pay the ultimate price for the months of hospitality she has enjoyed.

Nature has a terrible penchant for cheap thrills.

ANIMAL HUSBANDRY FOR DUMMIES

Domestic animals are always a challenge: Some creatures tend to turn up uninvited. Just think of dust mites, silverfish, and house spiders, all of which have found their niche in the corners of our homes. For ants, it is much the same. Aside from winter guests and parasitic gossamer-winged butterfly caterpillars who find their own ways into the nest, countless other guests can be found crawling about inside an ant nest in secret; the ants take no interest in them and sometimes they don't even know they're there.

Be it a human home or an ant nest, the benefits of life as a lodger are enormous: Inside, it's dry, warm, safe, and cozy, and, to top it off, a tasty morsel of food will drop in now and again. These harmless dining companions or commensals include the silverfish *Atelura formicaria*, which depends completely upon the food its hosts leave lying around as waste. Only when the hunger really kicks in does it pilfer a little of the bolus that the workers use to feed one another. Of course, this kind of thievery is not without risk; ants will make short work of thieves if they catch them. But like any silverfish, the *Atelura* is nimble and agile and can usually escape just in time.

EATING—AND NOT GETTING EATEN

The ants' less thrill-seeking guests take great pains to remain almost invisible to their hosts by, for example, holing up in unused corners of the nest, or by wearing thick armor, impenetrable to the ants' strong jaws. The ant bag beetle *Clytra laeviuscula* adopts this strategy in a particularly unseemly way by laying its eggs right at the entrance of the ants' nest and then defecating all over them. For some reason, the ants gather up the eggs and carry them into the nest, where the larvae soon coat themselves in a hard layer of excrement. This allows them to effortlessly withstand attacks from ants while also feeding not just on waste but also their hosts' offspring. An ant bag beetle larva will rampage through a nest for two years or more before it pupates and finally leaves as an adult beetle.

The most effective method for surviving inside an ants' nest, however, might be to bewitch the ants with the right perfume. The rove beetle *Lomechusa pubicollis* is an expert in this regard. To first of all find its way into the nest, it seeks an encounter with the common red ant *Myrmica rubra* and drums on one of the workers with its antennae. Once it has awakened

Some creatures are happy setting up camp in the lion's den: Guests of the ants, such as this beetle, live alongside them, where it's safe and there is always something to eat.

her interest, the beetle sticks out its bottom. This is equipped with a gland that produces a secretion that mollifies the ant, causing her to forget her aggression. From this point onward, the rove beetle need not worry about being eaten. Instead, it employs another trick: It presents the ant with the openings of two other glands on its bottom, which Bert Hölldobler has christened "adoption glands." Once the worker gets a whiff of these, she mistakes the beetle for a confused larva from her own colony and promptly carries it to the nesting chamber. It is then that this ungrateful fraudster goes to town, feeding on its hosts' eggs and larvae.

Life among a pugnacious colony of ants is not all that difficult once you get the hang of it. Over 3,000 species of springtails, beetles, spiders, woodlice, flies, and others know how. Some keep to the edge of the nest, like the flower chafer larvae. Others want to be at the heart of the action, perching on the workers themselves. Some mites have adapted to their very specific favorite spots, such as those of the genus *Macrocheles*, which cling exclusively to the outer edges of the feet of army ant soldiers. Their cousins of the genus *Circocylliba*, however, prefer to sit directly on their food source—that is, on their hosts' jaws.

Like humans, ants are surrounded by "pets" with whom they share their homes and their lives willingly, and they wouldn't want to be rid of them, even if they could be. But the worst parasites ants have to deal with are other ants, as we will see in the next chapter.

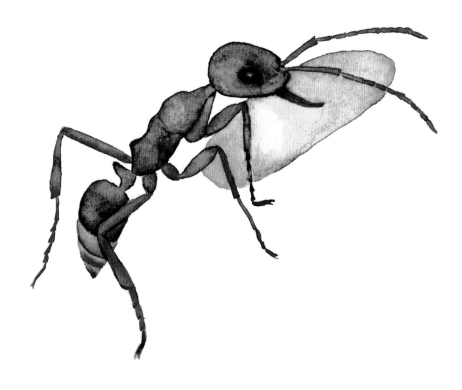

CHAPTER 10

ON PARASITES AND SLAVE-MAKERS

———

*A slave-making ant steals a
host's pupa on a slave raid.*

will never forget my first slave raid among the ants. While doing research for my master's and PhD, I worked with normal, nonparasitic ants, and so I was particularly excited to be able to spend my first postdoctoral placement in the US researching slave-making ants. I was hoping to investigate whether slave-makers and their hosts of specific regions had adapted to one another, so I positioned colonies from different regions in a laboratory arena to observe their confrontation. For example, New Yorker slave-makers against West Virginian hosts. Would slave-maker ants be more successful with host colonies from their region of origin, enslaving the host brood? Or had the hosts adapted to their local foes, and would they therefore be best placed to defend themselves against raids?

Instead of large wood ants, my test subjects were smaller ants, whose whole colonies fit inside an acorn. The benefit of this was that I could observe the entire slave raid from beginning to end in the laboratory. All I had to do was place the slave-maker nest, which fit between two slides, and a host nest in the arena, and wait for the slave-making ants to explore the area, discover the host nest, recruit additional comrades, form a raiding party, attack the host nest, steal the offspring, and carry them home. After that, all that was left was to calculate how many of the brood the slave-makers had carried away, how successfully the victims had defended themselves, and whether there were casualties or deaths.

That was my plan, anyway. Based on everything I had read, I was sure it would work. It would be easy as pie. Foolproof. But I may have overlooked one tiny detail. . . .

When I began my experiment in the first warm days of May, I was pretty confident. My slave-making ants were from the species *Temnothorax americanus*, from forests north of New York State, and would be accompanied by a colony of *Temnothorax longispinosus* from the same area, who would serve as hosts. I sat on the sidelines with a pen and notepad, ready to capture the slightest bit of action in detail. It could happen at any moment. Once a quarter hour had passed, I was sure things would get going soon. Then it was half an hour. An hour. Morning passed. Then afternoon. I sat perching on my chair until evening and . . . nothing happened. The slave-making ants spent the whole day sitting around in their nest doing nothing, not even poking an antenna out the door. The food scouts from the host colony explored the arena, but that was all. Disappointed, I went home and brooded over what I could have done wrong.

The next morning, everything changed. A quick glance told me that there had been an attack overnight. But the distribution of roles was not what I had expected. I hardly wanted to believe it, but during the hours of darkness, the workers from the host colony had marched out, forced their way into the slave-making ants' nest, and stolen their brood. What on earth was going on?

After a little research, it dawned on me that I had made a mistake in my experiment: My timing was all wrong. Slave-making ants are inactive almost all year round. They only steal from other colonies during raiding season. But raiding season starts in high summer, when the hosts' brood is more developed and ready to hatch, not in spring. In my zeal, I had scheduled the experiment too early, and the still-sluggish slave-making ants had been caught off guard. What I had not anticipated, however, was that the hosts would turn the tables on them. They clearly had the necessary disposition, since both species—slave-making and enslaved—are closely related; slave-making ants are simply a step ahead in their development.

At the end of July, the interactions had normalized at last. The slave-making ants grew suddenly restless, running out of their nest and back again, exploring the arena eagerly. They quickly discovered the host colony and raided them, as expected. I filled my notepad with observations. But I also learned a lot from the setback at the beginning of the experiment. It is often these little surprises that make science so exciting.

But ants are capable of exploiting each other without resorting to slavery, too.

GANG WARFARE

When ants attack their neighbors, it's often about food. Ants who find plenty of food can raise more offspring, lending their colony an advantage in the fight for survival. Since food is rare, however, and the search for food is exhausting, some species have found a more comfortable way to get at the loot: They simply steal workers from other colonies. Bert Hölldobler observed honeypot ants in the Arizona desert waylaying harvester ants in small groups, attacking them as they tried to carry a scavenged termite back to their nest. Though taken by surprise, the harvester ant was in no way defenseless and instantly went on the counterattack, but once it went to bite, it had to put down its prey, and the long-legged honeypots would quickly make off with it. In legal terms, this is a clear example of a heist, while it is known to biologists as kleptobiosis or kleptoparasitism. Ultimately, however, it amounts to the same thing: The harvester ant loses her termite and is forced to turn around on the home stretch and start her search for food all over again.

Though she has suffered a robbery, she should count herself lucky that her colony isn't visited by the thief ant *Solenopsis molesta*, which can be found throughout North America. Thief ant workers are some of the smallest ants native to these regions, measuring between one sixteenth and one eighth of an inch (1.5 to 3 mm) in length, but they turn this apparent handicap to their advantage on raids. Like bank robbers tunneling into a vault, thief ants build their nest close to that of their future victims. From this position, they build tunnels that are so narrow that workers of the other species are unable to pass through them, allowing them access to the nest and, ultimately, the nesting chamber. Whenever the little thieves are hungry, they set out in troops and grab whatever they fancy, entirely without asking. They are usually on the lookout for their unwilling donors' eggs, larvae, and pupae. If trouble strikes and the rightful residents spot the raid, the thieves spray a repellent substance from their venom glands, producing a similar effect to pepper spray and causing the host ants to reject their own brood. The thief ants split the stolen offspring into manageable pieces and schlep them back to their own nest. They often also keep a few underground aphids here as pets, feeding off their secretions. These small, yellow burglars and baby-killers don't depend on their criminal lifestyle to get by, but making a living by devious means is often easier—unless your victim has hired a couple of mercenaries. That's when things get risky, even for the slickest band of thieves.

THE MAGNIFICENT SEVEN

An unscrupulous band of crooks raids a village again and again, stealing its supplies, until the tormented farmers call for aid from a group of battle-hardened heroes. These plucky swashbucklers move into the village, are fed by the villagers, and live quite comfortably until the bandits reappear and the fight begins. *Seven Samurai* told this tale in 1954, before Yul Brynner took inspiration from it in 1960, fighting the same battle in *The Magnificent Seven*. But for ants of the genus *Sericomyrmex*, it's a story that has been playing out for hundreds of thousands of years.

These little ants put their heart and soul into farming. Like the better-known leafcutter ants, they carry plant material to their nests and use it to grow fungal cultures, which provide the food they themselves feed on. Their lives could be quite harmonious and peaceful if they didn't have to worry about attacks from raiding *Gnamptogenys hartmani*. These appear suddenly, killing all the *Sericomyrmex* workers in their path and getting stuck into the colony's larvae, without giving the fungal gardens a second glance. These bandits will destroy an entire colony in no time at all and then move on.

Unless the farmers themselves live alongside another species of ant. *Megalomyrmex symmetochus* lives happily alongside *Sericomyrmex* and is happy to mooch off its neighbors. Though the queen and her workers do not lift a finger when it comes to rearing the fungi, they feed off the fungal network and sometimes also munch on one or two of the farmers' larvae. They're quite a nuisance for ants tending to the fungus, but *Sericomyrmex* can't shake off its guests, because they bite the wings off the young host queens, preventing them from leaving for their nuptial flight and seeking new pastures. If ants could let out sorrowful little sighs, you can bet these unfortunate farmers would do so.

But their sacrifice pays off when *Gnamptogenys* appears at the nest door. While the farmer ants hole up in the depths of their hill, the lazy *Megalomyrmex* transform into heroic samurais and confront the enemy. They spray a toxic venom, felling not just the attackers but also countless warriors from their own ranks. Still, it's worth it. If these species are released at one another in the lab, it takes eight fungus-farmers but only two samurai to overpower an attacker. *Sericomyrmex* suffer few losses when their soldiers go to battle for them, while a colony without this kind of support will lose almost two thirds of its workers. Beaten and heavily decimated, the raiders are forced to retreat. They will try to target colonies without a hidden army of samurai in the future. Still, such a thing is hard to find because four out of five fungus-farming families harbors a secret clan of samurai. Just in case.

THE JOYS OF LIVING TOGETHER

The best kind of teamwork is when others do the hard work for you. In contrast to their hardworking image, many ants are extremely lazy. *Temnothorax minutissimus*, for instance, which I have found on the top of the Appalachian Mountains in West Virgina, is so work-shy that it no longer has a worker class. It only has queens and males (drones), a phenomenon we researchers refer to as inquilinism. For this kind of lifestyle to work, the inquilines need someone to keep a roof over their heads, bring them enough food, tidy up after them, and look after their offspring. In short, they need a residential community of sorts, one in which others do the work.

And what could be more natural than living alongside beloved relatives? In spring, having bred with their brothers the year before, the young queens move out to find a new home in the nest of a colony of *Temnothorax curvispinosus*. To avoid having their heads bitten off or getting thrown out as they attempt to sneak into their hosts' nest, these majestic little moochers imitate their hosts' chemical profile. This encourages the host colony to adopt them and integrate them into the community as additional queens. *Curvispinosus* workers tend to the *Minutissimus* queen's brood in such an exemplary way that a healthy scientific mind might ask why the host ants put up with this obvious display of social parasitism.

It's probably because the imposters are so well disguised; the exploited hosts have no idea that they are happily helping to rear a foreign colony of ants inside their own nest. After all, ants do not have many opportunities to check the identity of a given creature. This has a little to do with size and shape, but mainly, the creature must smell right to be accepted.

Inquiline species of ants, such as little Temnothorax minutissimus *(queens pictured at the bottom of image) have abandoned their worker caste altogether. Instead, their offspring is raised by the workers of their host species* Temnothorax curvispinosus *(large queen pictured above).*

Appearance does not pose much of an issue if the parasite is closely related to its hosts, as is almost always the case for socially parasitic species of ants. Smell is harder to mimic, but either the parasites have learned to copy their hosts' scent over the course of evolution, or they simply relinquish their inherent smell, allowing them to run freely under the olfactory radar in the nest, taking on the scent of the colony. When laziness is the order of the day, they are happy to leave their choice of deodorant to their hardworking hosts.

Inquilines also pay for their idleness with a greater form of disenfranchisement. They have no nest of their own, no territory, and are wholly dependent on their hosts. But nobody goes as far as the "last ant standing."

WHEN ANTS JUST WON'T LET GO

Not all remarkable ants require a journey halfway across the world, a fight through a mosquito-ridden jungle, and an ordeal by leeches as long as your finger. Sometimes, you need look no further than your own neighborhood. For me, that is the Western Alps. The French and Swiss parts of the Alps are home to an ant so rare that it wasn't discovered until 1949, and proof of its existence elsewhere has only emerged on occasion in the Pyrenees, Northern Spain, and Kazakhstan. Less than two dozen colonies have been discovered so far, making *Tetramorium inquilinum* a rarity among ants.

Three of these discoveries were made by the Luxembourgian naturalist Robert Stumper, who was goalkeeper for his national soccer team in his youth, proving that myrmecologists can be active aboveground, as well as under it. Searching for rare insects, he and a handful of colleagues dug out several hundreds of nests belonging to the pavement ant *Tetramorium caespitum* in the Upper Rhone Valley at an altitude of over 6,500 feet (2,000 m), before stumbling on three groups of parasitic *Tetramorium inquilinum*. One of the pavement ant colonies they encountered was discovered by accident when they lifted a large rock, which happened to roll down the slope and, on impact, split into two pieces along an old crack. They were astounded to see host ants and parasites swarming about on the newly fractured faces of the rock. The researchers collected a large sample of the insects and were able to transfer them to observation nests for use in the laboratory. A significant degree of our understanding of *Tetramorium inquilinum* can still be traced back to these colonies.

Like *Temnothorax minutissimus*, which we encountered in the previous section, *Tetramorium inquilinum* has also dispensed with the more demanding travails of daily life and does not produce a worker caste. Instead, it relies on its ability to exploit its hosts. The young queens have probably already copulated with the much rarer males in the nest. Afterward, many prefer to stick to the comfort of the mother nest, but some fly away to seek a new host colony. And they're pretty choosy. The only species they will accept other than *Tetramorium caespitum* is *Tetramorium impurum*, which is almost identical to the former. New research has shown that both potential host and parasite are closely related phylogenetically, meaning that they can both be traced back to a common ancestor, only becoming distinct species relatively recently. But relatedness alone does not suffice to ensure that *Caespitum* and *Impurum* workers will accept a young *Inquilinum* queen. Swiss and French pavement ants will tolerate a parasitic ant jumping onto their backs and letting itself be carried into the nest. The Luxembourgian colonies with which Stumper provided a couple of *Inquilinum* queens made quick work of these lonely moochers and killed them.

Once the *Inquilinum* queen has made her way into the nest, she tries to take the reins. Her dream is to live off one of the queens in the colony. She climbs onto a queen's back and clings to her upper thorax or abdomen using greatly enlarged claws. From here on in, it's a case of eating as much and as well as possible, growing as much as possible, and laying eggs. To ensure that the workers not only tolerate them but also tend to them, the insurgent queens produce an irresistible secretion that their on-duty maids lap up eagerly. In doing so, they both clean the *Inquilinum* and provide them with liquid nourishment. The parasites rely on these services because their withered mouth parts prevent them from cleaning themselves or biting into more solid substances. The rest of their bodies has also degenerated. Their exoskeletons are thin and pale, their abdomens are spoon-shaped to better conform to the shape of their steeds, and their brains are tiny, even for ants. Without the support of its host, *Tetramorium Inquilinum* would be lost; it would go extinct in a heartbeat, because pavement ants even take charge of rearing its young. This gives them plenty to do, too, because a mature parasitic queen produces great quantities of eggs in her bloated ovaries, averaging one every thirty seconds. So, she does everything she can, within the limits of what evolution has allowed her, to make sure her legacy lives on. The ancient Greek genus name for the *Inquilinum* is *Teleutomyrmex*, meaning, essentially, "the last ant."

Sitting astride her royal steed, a parasitic queen is rarely alone. *Inquilinum* has no problem sharing her host with fellow ants of the same species, so there may well be several ants perching on one queen. Stumper himself counted as many as eight *Inquilinum* sitting on a single queen pavement ant. These parasites did not kill their hostess outright, but the poor queen could barely move thanks to the crowd assembled on her body and staggered around clumsily, pulled this way and that by her own workers, losing a leg or two in the process, and ultimately perished. Tough as they are, even ants have their limits.

As far as we know, their well-adapted physiques and behavior make *Tetramorium inquilinum* the staunchest social parasites of the ant world. But they are not the worst pests. Ants can be much, much crueler.

WHEN RECRUITMENT GETS BRUTAL

Charles Darwin could hardly believe his eyes when observing the slave-making *Formica sanguinea*, the blood red ant. In *On the Origin of Species*, he writes about them and another slave-making ant, *Polyergus rufescens*, the European Amazon ant, arguing that "This ant is absolutely dependent on its slaves. . . . So utterly helpless are the masters, that when Huber shut up thirty of them without a slave, but with plenty of the food which they like best, and with their larvae and pupae to stimulate them to work, they did nothing; they could not even feed themselves, and many perished of hunger. Huber then introduced a single slave, and she instantly set to work, fed and saved the survivors; made some cells and tended the larvae, and put all to rights."

This description sounds much like the ineptitude typical of inquiline social parasites. But the European Amazon ant and other slave-making ants differ in two significant ways: They do have workers that in many situations are not put to work in the normal household. And they do not live with their hosts as lodgers, instead gathering their disenfranchised staff from slave raids on other colonies. For over fifty species of ant, this is a highly attractive way of making life easier for themselves. In fact, it is so attractive that ants have "invented" this slave-making habit at least nine times over the course of evolution.

THE FIGHT FOR THE HONEYPOTS

Honeypot ants are proof that slave-making, like thievery, often starts opportunistically. This is true for various species that employ the same methods to stock up on supplies when there is an excess of food. Special "storage" ants store sweet nectar in their crop, causing their abdomens to swell enormously. These little storage tanks dangle in the nest's "pantry" and cling to the ceiling, squeezing out a few nourishing droplets on demand. A storage ant is like a well-stocked freezer and is of great value to the colony.

The honeypot ants of the genus *Myrmecocystus* have another feature that sets them apart. If two colonies are close enough that their territories border one another, the conflict is settled with a ritual show fight. Both colonies send teams to the arena where the tournament is to take place, and the opponents take great pains to appear as large and as impressive as possible. They stand on the tips of their long legs, stick their heads and abdomens in the air and even climb up onto branches and grains of sand. Meanwhile, other members of the team crisscross the space, assessing which of the two sides

A specialized subcaste of honeypot ant stores supplies for the whole colony in its abdomen, causing it to swell. These "honey pots" are considered precious bounty by other raiding colonies.

seems superior. If the result is even remotely undecided, both sides will retreat peacefully. Imagine how much bloodshed could have been avoided had nations exhibited such behavior on the brink of World War I. In the honeypot ants, this entire procedure is repeated from time to time—the cool-down from one game is just a warm-up for the next.

But games can turn deadly serious when one colony sees it has the upper hand. The stronger colony is unlikely to turn down the opportunity to assert itself. Young colonies incapable of assembling large teams find themselves overrun, bitten to death, and robbed for good measure. The stronger group breaks into the nest and murders the queen and any workers who stand up to them. They then grab the eggs, larvae, pupae, and storage ants and take them to their own nest. If one of the less mobile honeypots dies on the journey, it is eaten, and the contents of its crop is later transferred to a storage worker in the victors' own colony. The survivors and the stolen brood are integrated into the victors' colony. Honeypots cannot run away independently, and the young workers, which later hatch out, see no reason to flee. They come to view the victors as their own colony and consider the other workers their sisters. They will soon adopt the scent of the nest and carry out all tasks required of ants of their age. Nobody would be able to tell them apart from the rest of the colony.

Unless they took a look at their genetic material. Though the workers of the winning colony smell like them and slave away like them, they are not their sisters, and the queen is not their mother. The pact that sees workers abandon their own offspring to give their siblings a better chance of survival and pass on their own genetic material indirectly does not apply to these stolen workers. Their genetic line ends when they die. And they are never rewarded for their hard work.

THE QUEEN IS A TROJAN HORSE

Enslaving ants from your own species is something of an exception. Slave-making ants usually take workers from other, frequently closely related species and install them in their own nests.

This slave-based economy generally begins with a newly mated queen who wants to establish her own colony. This shouldn't be too arduous, though it will be dangerous. Young queens on the search for a suitable throne force their way into the nest of a suitable host species. This is a particularly risky undertaking; if the colony's workers discover her plans, it will all be over for her in a matter of minutes. If the slave-making queen is lucky, she will come away with her life intact, though she will have been fairly battered. Sometimes, Calamity Jane will be missing an antenna or a leg, and occasionally several limbs may have gone astray. I once encountered an especially unlucky queen in New York: The head of a worker from the host species *Temnothorax longispinosus* was still clamped to her right foreleg. This appendage had probably once belonged to a worker defending its colony against a hostile takeover. In the struggle, the slave-making queen must have bitten its head off, causing the muscles to seize up in its jaw. Thus, the queen was forced to live with this grim millstone attached to her leg. I dread to think how the same scenario would play out in the human world.

To avoid a fight, a smart slave-making queen will carefully camouflage herself on a raid. A popular disguise is that of the "ghost ant," lacking its own scent. To achieve this, the slave-making queen must carry no or very little scent on her cuticle, so that the workers—who orient themselves mainly by smell—are dimly aware of her but cannot tell that she is an insurgent. Young queens of the same species spend a little time in nest's periphery before entering, adopting the scent of the colony. Others hotfoot it straight into the queen's chamber and assassinate her. When the slave-making *Temnothorax ravouxi* attack, the fight to the death can last days or weeks, during which the attackers appear to strangle the queen. However, ants do not breathe via an airway but via channels known as spiracles located all over their bodies, so the queen is not suffocated per se. She is probably slowly killed by damage to her central nerve cord. Amazingly, none of her daughters seem to care that their mother is being murdered. Not a single worker comes to the queen's aid. We still do not know why they're so apathetic. Presumably the slave-making ant prevents the dying queen from emitting a scent signal, or perhaps she gives out a mollifying pheromone masking the queen's cry for help, producing a kind of all clear.

Eventually, it's too late; the rightful queen is dead and the newly installed tyrant takes over the colony. The workers accept their new mistress without any sign of a protest. They feed her and raise her offspring. Instead of sisters, the nest is now home to newly hatched slave-making workers, who have no skill for housework or finding food. These tasks must continue to be performed by the workers of the host species.

The slave-making ant state has been established. For its neighbors, the long campaign of terror is just beginning.

PROFESSIONAL SLAVE-MAKERS

Slave-making workers are not workers in the true sense. They typically lack the instinct for everyday work like nest-building, caring for the brood, and finding food. They differ from their slaves not simply in terms of inborn behaviors, they are built differently: They are larger and have wider heads with strong jaw muscles and mandibles, which are dagger-shaped like a sabertooth tiger's or have sharp edges like wire cutters. In short, slave-maker workers look like fiercely armed fighters, a special unit whose mission in life is to pierce tanks and snap off legs and antennae. And when the number of enslaved ants in their colony drops too low, that is exactly what they do.

The first step in this violent recruitment campaign is to send out scouts to locate a suitable target for a raid. Scientists previously speculated that scouts preferred weaker colonies less able to defend themselves, making the risk to the slave-making ants as low as possible. We have since discovered, however, that the contrary is true. Since the risk of being killed on a raid is never zero, slave-making ants prefer to try their luck with fewer raids on large colonies promising plenty of slaves, rather than carrying out numerous attacks that yield little. Once a suitable colony of victims is found, the scout runs home leaving a scent trail. The fate of the chosen colony is sealed.

I witnessed my first slave raid in the wild—or "in the field," as we biologists say—in the Chiricahua Mountains in Arizona. I was there to collect other ants, but a colleague of mine was particularly excited about the North American slave-making ant *Polyergus breviceps*, so I accompanied her on a trip high up into the mountains. Thank goodness I did, because I was able to observe a stream of Amazon ants 8 inches wide and 30 to 40 yards (27 to 36 m) long, rushing along purposefully. We followed and eventually arrived at a nest of black *Formica* wood ants. The red Amazon ants were storming the host nest like a crawling inferno. Outside, all we could see were a few individual skirmishes, but we got the impression that the hosts were not really trying to defend themselves against the attack; they were more concerned with running away hectically and hiding as many of their brood as they could. The attack lasted several hours before the band of red Amazon ants retreated. Almost every one of the raiders was carrying a plundered pupa or larva in its mandibles. These would go on to become enslaved workers in the coming months.

Other slave-making ants, such as *Formica subintegra*, are less warmongering and a little more refined, but they are not any friendlier. Instead of fighting, they spray what it is known as a propaganda substance from their Dufour's gland, mimicking their victims' alarm pheromones and making them fight each other. These gland secretions can also have a calming effect, as we see with the "Ninja ant" *Temnothorax pilagens* from Michigan. Once sprayed with the substance, victims forget to defend themselves and even help their attackers to transport their brood to the slave-making ant nest. Sometimes, they even stay there to serve their new mistresses.

Whether the attacks are performed with brute force or perfidious chemical deception, the weakened colony will always need time to recover from the raid. Colonies do die after raids, but only occasionally. It's not in the slave-making ants' interests to destroy their victims, because they hope to be able to return next year.

However, the required number of enslaved ants cannot be met by a single host colony since enslaved workers make up by far the largest group in many slave-making ant colonies, and the slave-making ants' nests will not produce any workers of its own. In colonies of red Amazon ants, around 80 to 90 percent of the workers are enslaved. Myrmecologists once counted forty-one raids over the course of just thirty-three days one summer, in which a single colony carried off 40,000 pupae and larvae.

Little North American *Temnothorax americanus*, whose colonies fit inside hollow acorns, typically number thirty enslaved workers to every five slave-making ants. A third

If you want somebody to do your chores, you will have to find the right staff. The slightly larger slave-making ant Temnothorax americanus *is attacked by a colony of* Temnothorax longispinosus *on a slave raid.*

of all host nests within range are raided each year, meaning that practically no colony can develop as usual and the density of the nest drops to around half the usual number. When slave-making ants have free reign, they upset the balance in the whole area.

And they used to be normal ants, just like their neighbors.

IT'S ALL IN THE GENES

What turns hard-working ants into lazy slave-making ones? Human motives for depriving people of their rights and forcing them to work for us do not explain much when it comes to ants. Ants do not act out of greed or a lust for profit or power. They become slave-making ants simply because it works. And because their genes say so.

All ants were hardworking to begin with, providing for their own colonies. There were genes for practically every activity—from nest-building to caring for the brood, to defending the nest from enemies—and these governed the ants' behavior. Once one of the genes was lost, the ant affected by this had a problem: She no longer knew what to do, or how to do it. But luckily, she was part of a collective and one sister would step up to do what another could not. Unless, that is, the queen herself was afflicted by a genetic defect. In this situation, she would pass on this flaw to all her offspring, affecting the entire colony. From one generation to the next, this little branch of the species would, for example, forget how to feed its own larvae—and this would normally be a death sentence.

Unless, of course, the queen was in the habit of holing up in another nest and having her eggs tended to by its resident workers. We encountered this kind of social parasitism in chapter 2, The Birth of a Colony, in which the queen wood ants skipped the difficult start-up phase by unceremoniously capturing a ready-made nest together with all its workers. A few more changes must have taken place in their genetic makeup, however, for such a coup to succeed. The queen must have decided to seek out another ant's nest in the first place. Then, she would either have had to have no scent of her own, or taken on the scent of the desired colony, to avoid being bitten to pieces by the guards halfway into the nest. It would be handy, too, if the nest's defenders were lower down on the scale of aggression. And the list of conditions does not end there.

Developing a complex behavior such as social parasitism is no easy task and does not happen overnight. Most attempts fail because a necessary change is lacking or takes the process in the wrong direction. But evolution has had squillions of ant colonies at its fingertips, not to mention plenty of time. Now and then, different varieties combine to create a way of living that allows a queen and her colony to scrape by. When the queens and males of this new variety stop coupling up with insects from the original species—for instance, because they reach sexual maturity earlier in the year—a new species emerges. This new species resembles the original one in many ways, but its behavior can differ significantly. This stage sees many social parasites seek out shelter with close relatives. If they exploit several hosts at once, this can be the starting point for the development of a real slave-making species, which captures slaves from different, related species on raids. Alternatively, they might do away with a subcaste of workers and become inquilines, which typically specialize in exploiting a single species of hosts.

Genes remain at the heart of it all. In my lab, we look for the tiny changes that make a hard-working ant into a raiding slave-making one. New findings show that slave-making ants have lost more than a third of the genes governing the senses of taste and smell in comparison to their hosts. Since slave-making ants do not go out looking for food themselves and are fed instead, they get by with fairly limited senses.

But we don't just compare the letters in the gene sequences of different species. The genes themselves often do not differ per se, but their activity does. For instance, switching off the gene that governs caring for the brood is all it takes to make sure an ant has no idea what to do with a larva, even though it carries all the necessary information within itself. Other genes are far more active in slave-making ants. The *painless* gene, which controls pain sensitivity, at least in flies, is four times as active in slave-making ants. Some genes are involved in producing pheromones or have functions that play a role in the lives of slave-making in a way we have yet to discover. Compared to their hosts, the genetic variability of slave-making ants is decidedly high. This is to be expected because different species of *Temnothorax* have discovered slavery four out of five times independently of each other, and not always via the same path. What's more, some genes are activated or deactivated at different times, such that the relevant genes ensure that slave-making ants set out to find new workers during raiding season. These include the "clock" gene, which is known to regulate the circadian rhythms of animals. Interestingly, it has been shown that fewer genes are active during the raiding season than in the quieter months. Outside the season, the slave-making gene activity largely resembles that of their victims, aside for the inactive genes that prevent them from running their own households. Many slave-making ants flip-flop between personas, like Dr. Jekyll and Mr. Hyde.

But they are not the only ones capable of adapting to new situations thanks to their genes. Evolution has helped their victims to defend themselves, too.

THE VICTIMS' BLOODY REVENGE

Victimhood is no joke. It can even reveal hidden strengths. And for host ants, it has sometimes led to serendipitous genetic changes. For example, a queen can switch to fertilizing fewer eggs, producing more male offspring. In normal circumstances, this would be quite a foolish idea because males do not help develop the colony and their very existence can waste valuable resources. A colony with a surplus of males will usually quickly perish.

In regions where slave-making ants are wreaking havoc, however, other rules apply. It's not much use fighting the raiders, it's more sensible to flee—as far away as possible, i.e., by air. And flying is the preserve of queens and . . . you've guessed it: males. For a colony situated near slave-making ants, then, the presence of numerous flying males becomes a boon because they can carry the colony's genetic material far away where it might settle in peace. The colony may perish in a raid, but its genes will live on. The conditions turn a negative into a positive. Or, as we scientists might say, there is an evolutionary pressure to develop defensive measures against slave-making ants.

And that's where nature gets inventive.

Extra young queens also provide the colony with a better chance of escaping their nasty neighborhood during the nuptial flight. Flying reproductive animals make the colony smaller, but more mobile. Those left behind are often more capable of fleeing. Since there's little point in fighting the attacking hordes, the workers of some species don't even try; instead, they grab as many larvae and pupae as they can and set off on foot. In colonies occupying several nests, this strategy often results in the loss of only one of the nests and few of the offspring.

Another strategy reveals just how determined the hosts are to defend their nest. They limit the spread of their colony's scent by reducing the number of queens, each of whom contributes her own nuance to the scent. This more unified scent distinguishes the colony from other colonies, which is why slave-making ants, which mainly terrorize different host colonies, must select a particular chemical code in order to be able to trick their victims. When they bear down on their goal, they are met by workers that are better able to defend themselves than usual. *Temnothorax longispinosus* among others, increases its aggression when the slave-making *Temnothorax americanus* declares it is hunting season. Unusually for their species, several workers will descend on a raider at once, inflicting painful damage on the attacker through a coordinated defense. The beguiling propaganda pheromones sometimes fail to be of use here if, little by little, an increasing number of host colonies immune to manipulative scents prevails.

But these dramatic defense measures are a slaves' rebellion of sorts, occurring when it's already too late—and when no slave-making ant is counting on an insurrection.

SIX-LEGGED SPARTACUS

This is how it begins: The slave-making ants think everything is just peachy. They have plenty of ants they've enslaved, attending to everything with exemplary care, including tending to their mistresses' offspring. They clean the eggs regularly and feed the larvae until they pupate, before hatching out to become the new generation of slave-making ants. But it comes to nothing.

Out of the blue, the enslaved ants revolt, serving up a bloodbath among the pupae. They bite the defenseless young insects to death, rip them to shreds, and throw them out of the nesting chamber. Soon, two thirds of the pupae are dead. This is a heavy blow for the slave-making ants, who have not anticipated it and can do nothing to protect the brood.

As a researcher, two things interest me most about these bloody rebellions: What good does it do the enslaved to kill the pupae, when their colony was probably destroyed long ago? And why do they wait until the offspring have reached the pupal stage?

Maybe we can answer the second question by putting ourselves in the workers' shoes. They experience the world largely through smell. Ant eggs and larvae do not, however, produce a species-specific scent, so the workers remain under the illusion that they are caring for their own colony's brood. Pupae, on the other hand, produce a specialized cocktail of scents. This prompts the enslaved ant to realize she is dealing with a foreign species of ant and her instinctive urge to fight the enemy kicks in—with fatal consequences for the pupae.

Enslaved workers do not personally benefit from this act of sabotage, of course. Yet their rebellion can be of some indirect benefit, because it weakens the slave-making colony, leaving it with fewer ants to send out on raids. If the enslaved worker's colony has not been completely destroyed, however, it gives her sisters valuable time to get back on their feet. What's more, previously formed sister colonies in the neighborhood are granted a grace period in which they are not subject to attacks. By killing the next generation of slave-making ants, the enslaved's rebellion increases the likelihood that her sisters, nieces, and cousins will survive. And when it comes to the struggle for survival, small victories can be crucial.

Where slave-making raids occur, there is an evolutionary arms race between slave-making ants and enslaved ants. Both groups exhibit a larger genetic range compared with normal colonies, because different changes, which otherwise would not have stood a chance, have prevailed, creating a distinct advantage. Each population develops its own strategy. But scientists have observed that colonies under pressure tend to produce more multifunctional individuals and fewer specialists. When life gets hard, it helps to know the babysitter is plucky enough to pitch in elsewhere when required.

THE ENSLAVED, BUT NOT AS WE KNOW THEM

We have been looking so closely at the enslaved and the slave-making that I almost can't bring myself to admit that what we observe in ants is not true slavery in the strictest sense. The term only really goes some way to describing this phenomenon. Admittedly, many of the characteristics of slavery do apply, such as the abduction and exploitation of a labor force, but a crucial aspect is missing: Where we see slavery in human societies, it is always fellow members of the species who are subjugated. In this respect, what intraspecific social parasites such as honeypot ants do is indeed slavery, but the actions of the raiding slave-making ants and *Temnothorax* are much more akin to what humans have done in domesticating animals. Humans captured free-roaming cattle, locked them away, and drank their milk. To this day, horses are forced to toil for humans, and pigs end up on the dinner table. Similarly, slave-making ants source their workforce from colonies of different species of ants—though that doesn't make it any less brutal.

Nevertheless, this habit of speaking in terms of slave-making has gained currency among scientists. The ants themselves don't much care what we call them. All they care about is getting their genes over the next hurdle in the struggle for survival.

CHAPTER 11

PHYSICIAN, HEAL THYSELF

Nematodes use ants as intermediate hosts. In the hope of attracting birds as additional hosts, they reprogram the infested ant's metabolism, so the insect develops a red abdomen and looks like a ripe berry.

H ere's where it gets a bit gruesome. In this chapter, we'll be looking at diseases that afflict ants and they're a little unappetizing, to say the least. We'll also encounter real-life zombies, which crawl around in the ants' remote-controlled remains. But there is a happy ending—I promise—because evolution has helped the victims to defend themselves. So, join me as we enter the hospital ward of the ant world.

BOILS, PIMPLES, AND LUMPS

It's not all that easy to tell when an ant is sick. Ants are anything but fussy and will plough on as usual when they've lost a leg or an antenna. After fighting another colony, some of them will carry around the severed head of an enemy still attached to their legs. Trifling matters like a cold or an upset stomach—if insects can even suffer from such things—would not deter any ant from scrupulously carrying out her duties, though some viruses can strike them down. But in these situations, the sisters will usually discreetly dispose of the body on the colony's own garbage dump to limit the risk of infection before we myrmecologists notice anything.

We only realize something's not right when our guests begin changing shape or color—for instance, when the colony is afflicted by mites. These arachnids are plainly smaller than their hosts, but an ant covered in dozens of mites looks battered—as if it's been through a heavy hailstorm—and its mobility is affected, to the extent that it eventually starves to death. When boils and pimples spill out of the body, this is generally a sign of some fungal infections—and at this stage, the ant is already dead. Sometimes, the fungal fibers eventually overrun the whole body or billow out of the ant's mouth and joints until it looks like a stuffed animal.

It gets really nasty when nematodes move in. Some species roam an area of the abdomen known as the gaster, eventually growing too large to fit inside the insect. In such situations, hobby breeders of ants are often initially pleased to see how well-fed their ants are looking, but in reality a nematode nearly 3 inches (7 cm) long has squeezed itself inside the ant's abdomen, itself only a quarter of an inch (7 mm) in length. The vital organs are squashed. But nematodes have an even uglier ploy to cheat their way through life on the ants' dime.

FALSE FRUITS

There's famously no accounting for taste, and there's no reason why we humans should like what ants do. Bird poop, for instance. While we might grumble at the sight of an ugly-looking stain on our jackets, cars, or even monuments, ants are just happy to get a free portion of minerals and nutrients. They gather up the remains eagerly and take them back to the nest to feed them to their larvae. And that's where the drama begins.

In Central and South America, bird poop sometimes contains the eggs of the nematode *Myrmeconema neotropicum*. A little later, when the larva, which ate the poop, develops into an ant and hatches out, the young worms become active in the ant's abdomen. This provides the conditions every parasite craves. In these sheltered surroundings, the nematodes mate and lay their eggs. And then they have just one more problem to solve: Life inside the ant may be comfortable, but you don't get far when you are on six short little legs. If the parasites are to spread over a larger area and find new hosts, their eggs must find their way back inside a bird, which they employ as an air taxi of sorts. The problem is that the worms' favorite birds don't eat ants, preferring colorful berries instead.

The nematodes' solution is as simple as it is incredible: They turn black ants into red berries. Exuding a specific blend of semiochemicals, which interferes with the ant's metabolism, they make the ant's abdomen swell and turn red. These substances also slow down the otherwise lively little ant until it can hardly move. To top it off, the parasite forces the ant to stick its rear end in the air, as if in alarm, and soon there's little to differentiate it from a real berry in the eyes of a hungry bird. A quick peck and the ant is dead and the nematode has achieved its goal. The journey through the bird's stomach and intestines do no damage to the worm's eggs. Unharmed, they await their next opportunity to be expelled and, with a little luck, become a tasty snack in the belly of the next ant larva. And so, the cycle continues.

Altering the body of a living ant? Doctor Frankenstein would be jealous to know how easy it is for these simple little nematodes. And he'd be even more surprised to discover what tapeworms do.

LONG-LIVED AND WORLD-WEARY

The tapeworm's strategy is similar to that of the nematode. They, too, find their way into ant larvae as eggs in the feces of birds, bore through the ants' intestinal walls, and establish themselves in the abdominal cavity. An infestation of the tapeworm *Anomotaenia brevis* in the ant *Temnothorax nylanderi* can be identified by a different body coloration: Instead of the normal light brown, the affected insect will emerge from the pupae in light yellow. While I was completing my PhD at the University of Würzburg, I was keen to find out what the parasite achieved by interacting with the ants, but unfortunately the local colonies were only very rarely afflicted with tapeworms. So, once I transferred to the University of Mainz, I was delighted to discover almost one in every three colonies here contained yellow ants. The ants themselves are naturally

When an ant larva is afflicted by a tapeworm, the resulting ant worker develops a light yellow color instead of brown. Its life expectancy also increases many times over.

a little less pleased about it, but they gave my work group the material necessary to prove that tapeworms wreak havoc not just in the lives and bodies of the afflicted ants; they also manipulate ants that are not afflicted. And they have it in their power to extend the ants' lives—or cut them short.

We decided to focus on *Temnothorax* because they are easy to gather and whole colonies can be kept in laboratories. This was necessary to see whether the colony demonstrated a reaction to the parasitic infestation. We collected huge quantities of acorns from the woods and inspected them for ants and tapeworm-infested colonies. This left us well-equipped for a whole range of experiments— with a somewhat surprising result.

We discovered that ants are very aware of which workers are carrying tapeworms and which are not. They determine this by smell: Sick workers smell different because the composition of chemical substances on their cuticles changes in response to the parasites. But instead of taking the precaution of throwing their sick sisters out and ridding themselves of the tapeworms, the workers did the exact opposite: They tended to the sick ants quite movingly, bringing them more food and cleaning them regularly. In fact, they cared for the afflicted workers better than they did their own queen. This had two consequences. Their care prevented the sick ants from getting hungry or catching a fungal infection. And yet, the rest of the colony had to go without the additional food rations and the time spent tending to the sick ants. In small colonies comprising just a few dozen insects, this seemingly tiny loss has tangible consequences. When compared to colonies without tapeworms, healthy workers and queen in nests with one or several afflicted ants lived considerably shorter lives.

The difference between those carrying the tapeworms and their unafflicted sisters was even starker. Those ants with a tapeworm in their abdomens lived ten times longer than the healthy ants in their colony. Their life expectancies grew to be on a par with that of a queen, which—in this species— can live up to twenty years. The afflicted workers were also able to develop their ovaries and lay eggs if the queen was removed from the colony. This begs the question: Are tapeworms a kind of fountain of youth and fertility elixir for ants? How in the world could tapeworms, of all things, perform such miracles?

THAT DARN TAPEWORM!

As is so often the case, the answer lies in the genes. When we compared the genetic activity of afflicted and unafflicted workers, we encountered around 400 variations. Many of these controlled the locomotor system. By somehow ensuring that these genes are switched off, the parasite impedes the development and function of the muscles, which is why the ants struggle to move and cannot run away when danger approaches. This makes them easy prey for predators such as woodpeckers, which is exactly what the tapeworm wants.

Other genes work to extend the lifespan of the infected worker. The genes for synthesizing new proteins and combating oxidative stress are much more active in ants carrying a parasite. They are probably similarly affected by proteins that the tapeworm produces in the ant's abdomen. We cannot currently say precisely how this works to extend the ant's lifespan. All we know for sure is that, once again, everything works in the tapeworm's favor: The longer an afflicted ant lives, the greater the chance of its being eaten by a woodpecker and completing the tapeworm's cycle.

But the tapeworm is not content simply to tamper with the genetic activity of its immediate host; it also manipulates the gene activity of the healthy ants in the colony. It interferes with the balance of the messenger chemical *tachykinin*, which is integral to the development of aggressive behavior in many species of animals. Indeed, ants whose colonies contain afflicted insects are less aggressive toward ants of the same species from different colonies than those from colonies free of tapeworms. It's a somewhat stupid development for this now peaceable colony, because brutal competition reigns between different colonies of *Temnothorax* over the few suitable nesting sites, and whoever is too compliant will soon find themselves flying off and away from their cozy nest. Tapeworms, however, are much keener on gentler ants. Only once a colony is less meticulous about the scent of its own nest and begins to accept different-smelling ants of the same species will the tapeworm's host worker be able to pass within the colony with its different scent. Otherwise, a naturally aggressive colony would, in all likelihood, bite its sick worker to death.

At the end of the day, a tapeworm reprograms not only the unlucky ant in which it has made its home, it reprograms the entire colony. It's a tyrannical dictator, pulling its hosts' genetic strings, making them do everything in their power to ensure that the worm finds its way back into the body of a bird. It does not give two hoots about the ant colony. What a jerk!

Still, it gets a whole lot worse.

REMOTE-CONTROLLED DEATH WISH

Sometimes my job as a researcher allows me insights into the behavior of my fellow humans that are almost as fascinating as what I encounter in ants. Say I am at a party and someone asks me what my job is: I could bet a lot of money on the follow-up question having something to do with tackling ants at home or in the garden. Fair enough (I've provided a couple of tips in chapter 12, The Path to World Domination), but this really isn't my area at all. Instead, I mostly tell people about what happens on a slave-making ant raid, or that some parasites turn ants into zombies devoid of free will, ordering them around however they like. This seems to work. Suddenly everyone seems to forget the ruined picnics and poisoned bait and wants to learn more about the zombies.

Of the many kinds of ant zombies, the tropical carpenter ant *Camponotus leonardi* has been the subject of the most research. These insects settle primarily in the treetops, presumably because they are safer there from one of their treacherous foes: the fungus *Ophiocordyceps unilateralis*. Nevertheless, if a carpenter ant catches a fungal spore, the fungus threads, known as hyphae, work their way into the insect's body and take control. Contrary to previous assumptions, they do not take control of the brain to do this. Instead, they wheedle their way straight into the muscle cells. In half a week or so, they have spread to such an extent that they are able to control the ant's behavior. They use messenger substances to give the muscles commands that the ant is powerless to resist, though it presumably experiences it all up close. Little by little, the fungus forces the worker to crawl down from its nest tree to where it is warmer and more moist. The parasite is most likely to thrive where humidity is over 90 percent and at temperatures between 70 and 90°F (20 to 30°C). It keeps the ant about two hands' widths above the ground, then makes it tramp sideways around the trunk of the tree. Its aim is to reach a leaf on the north side of the trunk. Once there, the remote-controlled ant bites down hard on the leaf's main vein. In doing so, it achieves the first part of the fungus's task and can then die.

The second part sees the ant's body serve as fertilizer for the fungus. The fungus grows, and after about a week, its threads break through the hard cuticle on the ant's head and form a stalk twice the length of the ant's body. In two to three weeks, fungal spores develop on this "horn," eventually raining down onto the ground. From their slightly elevated starting point, the spores can cover an area of around one square yard, much further than they could from the ground. Once there, they wait for a new ant to capture and force to capitulate to their demands. On some trees, the fungus affects so many ants that the trunks become veritable ant graveyards, each square yard host to dozens of zombies hanging from the leaves.

This strategy seems to work well because the fungus has been turning ants into mindless transport and food receptacles for at least 48 million years. This was the estimated age of the bite marks found in fossilized leaf remnants discovered in the Messel Pit, investigated by the Bonn-based paleontologist Torsten Wappler. The veins of these leaves were surrounded by tiny holes, which perfectly matched the imprints of ants.

PART-TIME ZOMBIES

But zombies don't necessarily find themselves steered toward leaves in moist, warm environments. Another fungus, *Pandora myrmecophaga*, forces its victim to scale stalks of grass and bite onto them. This is intended to scatter the spores further from an elevated point.

The lancet liver fluke *Dicrocoelium dendriticum*, on the other hand, has a different goal in sight. As far as it's concerned, the ant is just an intermediary host in a highly complex life cycle. It mainly lives on herbivorous mammals such as rabbits and deer, but can also be found in sheep, goats, cows, and horses. It has even been documented in dogs, pigs, and rodents, occurring occasionally in birds and, incredibly, humans. The liver fluke's eggs find their way into pastures in feces, where they are then eaten by snails. Here, the parasites change form and, wrapped in small balls of slime, set out once again on their journey. These slimeballs are considered a tasty treat by ants, which the liver fluke turns into zombies. Once they've eaten them, the afflicted ants begin to feel an urge to scale vegetation, flowers, or grasses, where a passing grazing animal will hungrily hoover up the ant and the parasite as part of its dinner.

Sometimes, however, neither cows nor rabbits will materialize, and the ant will wait in vain to be swallowed up. The liver fluke has expanded its zombie-making tactics for situations such as these. As long as the temperature remains under 60°F (15°C), as usually occurs from evening through to the following morning, the ant's mandibles remain firmly clamped on the leaf, flower, or stem. If the ant is not eaten during this time and the temperature begins to rise over the course of the morning, its jaws will loosen, and the parasite will let the ant go about its life as usual. It will crawl back into its nest, where it will go about the normal day-to-day life of a worker in its colony—until the sun sets and the temperature drops. Then, the zombie will be drawn back out into the open, where it will offer itself up as an evening meal.

Fungi turn ants into zombies. They grow around the ant's muscles and force them to seek out suitable spots from which the fungus can spread its spores.

VACCINATION OR DEATH

Ants must shoulder some of the blame for all this misery with fungus and worms: They live crammed together in nests full of microorganisms of all kinds, at a fairly constant temperature and humidity, and are always nibbling away on dead or live prey, often packed with viruses, bacteria, and fungi. To top it off, members of a colony are so genetically similar that any pathogen that infects one worker can go on to afflict them all. It would hardly be surprising, you might think, if they all got sick.

But they don't.

Ants have developed a sophisticated multilevel defense system to prevent one worker's infection from wiping out the whole colony. This boasts amazing parallels to our own immune systems, which is why it is known as the social immune system.

The first step is to prevent pathogens from getting in. While our skin performs this for us, the same job is carried out by the workers guarding the entrances to the nest. They thoroughly clean ants returning from foraging trips and remove the majority of harmful bacteria and fungal spores. In doing so, they are obliged to come into contact with all kinds of pathogens, though these are in comparably small quantities— not enough to cause sickness, but enough to enable the ant's body to prepare to defend itself against them. If it encounters them later in a more concentrated form, its body will respond accordingly; it may only get slightly sick, or it may not get sick at all. This cleansing process is essentially a simple form of inoculation, much like that which we humans once used against smallpox, known as variolation. People would inhale smallpox pustules ground into powder or scratch a little of the liquid from smallpox blisters into their skin, causing a minor case of the infection, which prepared the immune system for a proper confrontation with the virus. The vaccinated person would usually then be protected against the illness but would sometimes become so sick that he ultimately died. This is why variolation was superseded when safer methods of vaccination emerged.

Sometimes cleaning is not enough and a worker will still get sick. However, before the sickness emerges properly, the ant's body will often realize something is wrong and send out a warning signal to other members of the colony. As with tapeworm, sickness will alter the composition of hydrocarbons on the ant's cuticle, so that any ant encountering its sickly sister will know. The affected ant may also release a pheromone to inform more distant sections of the colony, but this is still speculation. The affected ant is careful to stay away from the nesting chamber, because nothing would be worse for the colony than if the offspring were to fall ill. Thus, the colony's entire social network changes in response to the chemical signal from the infected ant. The queen limits contact to selected members of her sheltered royal household.

The ants tasked with caring work, however, try to treat the sick worker with additional cleaning and all kinds of pharmacological tricks, which we will examine more closely in the next section. This allows for a kind of social vaccination to spread through the nest, with increasing numbers of insects becoming better protected. But if none of this works, the colony is left with one option to avoid catastrophe: The sick sister must go.

In beehives, situations such as these are occasions of great drama, with workers violently ejecting the infected sister. For ants, however, the process is often a quiet and secretive affair. It often comes from the sick ant herself. At a certain point, the ant begins avoiding contact with her sisters and withdraws. She no longer shares food with the other workers, ducks out of her caring duties, and ultimately leaves the nest. Outside, she looks for a spot that will be of no interest to the colony's workers on foraging duty, where she will not disturb anyone. Here, she waits for death. The fact that sick ants mostly choose to leave the nest to die, rather than being forced to by a fungus or another pathogen, was proved by researchers in an experiment in which they sought to almost suffocate the ants. The doomed creatures wandered breathlessly into exile of their own free will. They even moved out a second time when the scientists dispatched them back into the nest with tweezers. An ant will place the well-being of its colony above its own life—surely the stuff of Hollywood blockbusters.

But it doesn't have to end in heroic self-sacrifice. A well-stocked pharmacy will help fend off many furtive attacks, and this is how ants defend themselves against fungi and other parasites.

INSIDE THE ANTS' MEDICINE CABINET

Marveling at his colonies of moldering bacteria in 1928, the Scottish doctor and microbiologist Alexander Fleming was not the first to be interested in the beneficial effects of antibiotics. In 1910, his German colleague Paul Ehrlich successfully treated a case of syphilis with antibiotics, and fifteen years earlier, the French physician Ernest Duchesne had discovered that Arab stable boys allowed mold to grow on saddles, as it helped the horses' saddle sores heal quicker. But even he was light years behind the ant world.

In chapter 7, A Garden for a City of Millions, we saw how leafcutter ants employ antibiotics to protect their fungus gardens against the parasitic fungus *Escovopsis*. What's astounding about this is not just that the insects happened upon the idea of forming an alliance against pests with antibiotic-producing bacteria in the first place, it's also that the fungus has not succeeded—over 50 million years—in developing a resistance to this substance. In contrast, when we humans place a new drug on the market, it's only a few years before we have to grapple with pathogens invulnerable to our sup-posedly magic bullet. Scientists from the US and Costa Rica wondered whether ants and their antibiotic-producers of the genus *Streptomyces* were simply better at it. They isolated a component in the mix of substances used by the leafcutters to treat their gardens, known as selvamicin, and even applied for a patent for it. Whether it will make it into our pharma-cies remains to be seen. Selvamicin seems to be effective, at least against the nasty yeast fungus *Candida albicans*, which thrives on practically all human skin and mucus membranes, and inside our digestive tracts.

And selvamicin is not the only attempt to find something that might benefit human medicine in the ants' medicine cab-inet. The Kenyan ant *Tetraponera penzigi*, which lives in hol-low chambers in acacia trees, also uses antibiotics to protect its fungus gardens. A specimen killed off—at least in labora-tory tests—the much-feared strains of methicillin-resistant *Staphylococcus aureus* (MRSA) and vancomycin-resistant enterococci (VRE). Both pathogens cause great difficulties in intensive care units and nursing facilities because they cannot be combatted with any traditional means. A little help from the ants would be more than welcome here.

What is it advertisements say? "Side effects may include…" And side effects sometimes emerge where we would least ex-pect them. For instance, the Australian insect researcher An-drew Beattie determined that the antibiotics that ants use to fight pathogens inside their own bodies also kill the pollen of many plants. Often all it takes is the slightest touch. Perhaps that's why ants are such lousy pollinators when compared to their cousins, wasps and bees. Even if they tried, they wouldn't be doing the flowers any favors. Apart, of course, from a couple of exceptions: The flying males of *Myrmecia urens* of Southern Australia pollinate an orchid and have even developed a mo-nopoly on this service. To pull off this trick, the flowers use a specialized secretion to protect their pollen from direct contact with the flying ants. Other pollinating ants reduce their production of antibiotics during pollen season. Let no one say that ants can't do whatever bees can do.

But it does not always have to be antibiotics when it comes to keeping the nest and the ants' bodies free from pathogens. The wood ant *Formica paralugubris* achieves the same thing with little clumps of pine resin, which it collects and carries into its nest. Twice as many workers and larvae survive infections with bacteria or fungi when compared with colonies without access to resin. We still do not know which of the countless components of resin is responsible for this protective quality. We lack the ants' experience, and they have been experts in natural medicine for at least 50 million years.

Particularly, it turns out, on the battlefield.

KILLER COMMANDOS—AND NURSES

The going gets pretty tough for ants. Especially for species that specialize in attacking their well-fortified contemporaries and have to reckon with their victims biting back angrily, leaving one or other of the sisters to fall by the wayside. A colony of millions brushes off such losses, but when a colony only numbers a few hundred insects, every fighter counts. That's why some ants tend to their wounded so carefully.

Megaponera analis, for instance, is nearly three quarters of an inch (2 cm) long, making it one of the largest species of ant in the world, but when it comes to reproduction, these ants leave much to be desired. Just around a dozen new workers emerge on average per day to see the light of sub-Saharan Africa. With so few troops to speak of, a 2,000-strong colony of these ants would be considered large, and they are usually far fewer in number. All the crazier then, that *Megaponera* prey specifically on termites of all creatures, and have a big appetite to boot. Several times a day, troops of 200 to 600 fighters will

When attacking a colony of another species, raiders from the species Megaponera analis *do not abandon their lightly wounded sisters; they carry them back to the nest for treatment.*

set out to storm a termite's nest. Large majors rip through the nest walls using their strong jaws, clearing the way for nimble minors. Meanwhile, the termites dispatch to the site of the break-in a caste of soldiers developed especially to defend against *Megaponera*. War rages inside the nest; the fighting ants often lose a few legs, or find termites clamping onto them with their jaws. Both prove to be big obstacles when the raiders try to retreat with their prey, and workers wounded in this way will often not be fast enough to keep up with their sisters.

Observing these situations while studying for his PhD at the University of Würzburg, Erik Frank encountered a behavior that is unique to these little invertebrates. The raiding party has injured workers stick close to the path home and call for aid from their comrades, depending on the species. They send out two pungent pheromones, which smell like rotten eggs or stinky cheese. Their fellow raiders are not put off by the smell, however, picking up the weary soldiers and carrying them home to the nest. Once they arrive, the nurses set to work tending the wounded. They carefully nibble off any termites that have attached themselves to the patient and lick her wounds. Researchers have yet to determine whether this is when they administer antibiotics. Either way, the treatment is a success: While one in three workers to whom scientists forcibly refused treatment died within a few hours, nearly all the ants permitted to receive care from the ant nurses survived. And since around a third of all ants in a colony will lose one or several of their legs during their lifetime, the nurses' medical care improves the colony's headcount by about a third.

Does this mean *Megaponera analis* has its very own Red Cross, tending selflessly to its sick and wounded? Not at all. As we often see with the amazing abilities of ants, the insects are unknowingly conducting a coldhearted cost-benefit calculation as they tend to their comrades. Is it worth it for the colony to bring an injured worker home? Or will she be a burden on the colony? The decision is pretty simple: If the ant is merely handicapped by a clinging termite, or has lost a couple of legs, she will either be fit for action once the foreign body is removed or will become accustomed to her missing limbs in a few days and be able to run almost as fast as before. If, however, three or more of her legs have been bitten off, she will never recover enough to be able to go raiding. She is better off dying on the battlefield. This might be a fairly gruesome decision from our perspective, but for the ants it's a logical conclusion, made by the injured party herself: Badly injured workers do not send out chemical calls for aid, they simply surrender themselves quietly to their fate. These ladies are true warriors to the last breath.

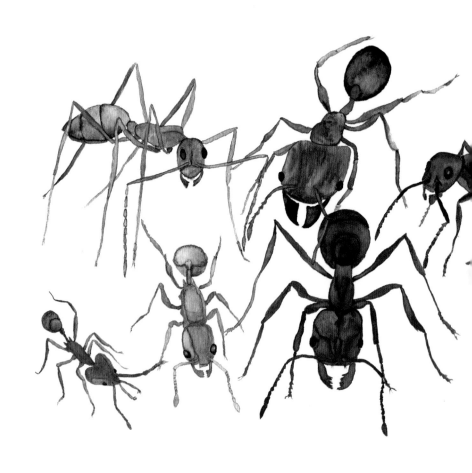

CHAPTER 12

THE PATH TO WORLD DOMINATION

All kinds of species of invasive ants have embarked on a quest for world domination.

The conquest of the New World began in the sixteenth century. It was around this time that Spanish sailors tore a little family from their Mexican home of Acapulco. They hauled them over the Pacific Ocean to Manila in the Philippines, buried in several hundredweights of earth. At the time, the city was one of the most important trading points with the Far East and the Spaniards found profitable cargo there. They threw the redundant earth on empty areas of the harbor, and the uprooted creatures went with them. They had no idea they were helping the little family establish a new empire, which would one day encompass the globe.

Today, we find the descendants of this founding population of tropical fire ants *Solenopsis geminata* in all kinds of places where they don't belong. From Manila, the sailors took them to China and different trading points across Southeast Asia. On the journey to Europe they also found their way to Africa and they accompanied the British settlers to Australia, too. Genes and shipping lanes uncover the stories of the past. But they also reveal something else: Though these ants are scattered across the globe, they all belong to one family. An ant from Portugal could march straight into a nest in Australia and be greeted like a sister. They are part of a supercolony that has conquered the world by force. There are more of these global mega-families than you would think.

And some of them make life a living hell for humans.

ANTS AT HOME

In their home country, these little conquistadors are often normal ants. Most species live in colonies with one or a few nests and one or a few queens. Their kingdoms encompass a tree or a garden, and wood ants' domains can even cover whole forests. However, their territories rarely stretch beyond this. This is due to constant disputes with their neighbors, who compete with them for food, aphids, and nesting sites. This proliferation of small states keeps them from getting a taste for greater power.

And then there are the problems ant colonies grapple with, as we have seen in this book. Fungi and other pathogens, which can—in extreme cases—wipe out whole colonies. Birds, lizards, spiders, centipedes, and other predators, for whom the nuptial flights signal the start of a great feast. It's a well-practiced arrangement in which the players work together and against each other to ensure that everything stays within its limits and no one gets the upper hand. For the ants, it is a daily struggle to survive; for the ecosystem, it's a dynamic equilibrium; and to us, it looks like paradise.

Until humans intervene and carry off part of this kingdom to a different world.

ALL ALONE IN A FOREIGN LAND

It's usually accidental. All it takes is for a single queen who has recently mated to find her way into a pack of tropical fruit during harvesting or for a small portion of a colony to crawl onto a ship, as in the case of the tropical fire ants. Ants will often find themselves involuntarily relocated when a botanic garden or a park orders in an exotic plant. It doesn't even have to be a living tree; ant colonies can make their homes in tropical timber, too. Your average tourist may turn ant smuggler when he or she unknowingly bring home a souvenir that turns out to be occupied. And let's not forget that some nations have imported foreign species in the (misguided) hope of tackling local pests. Worldwide trade and globalization don't just carry goods and people across the world, they carry ants, too. And yet, while people know what to expect, ants find themselves dropped in unfamiliar surroundings and have to work out how to make the best of it. In most cases, the experiment goes awry and the travel-weary ants perish.

But sometimes they embark upon a conquest of discovery, such as the country hasn't seen in many years.

To do this, they must first become established, i.e., survive. Their country of exile can't be too hot or too cold, nor too dry or too damp, and there must be suitable food as well as appropriate nesting sites. If these conditions are met, the resettled queen has a chance of establishing a first colony. If it all works out, her colony will thrive and grow quickly. Sometimes this will happen faster than would have been possible in the ants' homeland.

Sometimes, it's true that there's magic in new beginnings. If the transported ants were healthy in their previous home, they will remain so in their new home, because there are often no pathogens that might infect them. Their new home will often also be devoid of specialized predators. And there may be no ants of the same species to compete over territory. In short, it's paradise there for the taking. And some species of ant don't need to be asked twice.

WE HAVE COME TO CONQUER

Spain no longer belongs to the Spanish, nor France to the French, and the locals have lost control of Portugal. The entire northwest coast of the Mediterranean Sea and the Atlantic Coast of Gibraltar as far as the border between Portugal and Spain is firmly in the hands of the Argentine ant *Linepithema humile*. These ants dominate the ecosystem across an area covering 3,700 miles (6000 km). Where was once a seething mass of different species, there is now far-reaching monotony, because there are few or no colonies of indigenous ants left. Even the conquerors themselves cut a fairly drab image. You see, they haven't established several different colonies and spread; instead, just two colonies reign over the entire coastal area. The west of Southern Europe finds itself in the hands of two mega-families with umpteen billion members. And these ants exhibit family relationships that the Mafia can only dream of.

It all began in 1895, when a queen was brought to the area and succeeded in founding a colony. She didn't seem much of a threat to the local ants, because the Argentine ant is decidedly small. The workers only reach up to one eighth of an inch (2 to 3 mm) in length and the queen is around twice this size. Not only that, she was all alone: the young queens and males from her first brood could not find partners from other colonies. The royal brothers and sisters had no choice but to mate with each other. They stopped flying out, instead copulating inside the mother nest. The daughter-queens then remained at home as co-regents. When the number of insects in a colony grew too high, one or several of the queens would move out along with a portion of the workers and offspring and form a new nest that remained so closely connected to its mother nest that the ants were at home anywhere and would come and go as they wanted, unchallenged by the other ants. Essentially, this was still a single colony, now inhabiting several nests, a phenomenon which we myrmecologists refer to as "polydomy."

And so the Argentine ant's territory continued to grow. One nest became several, several became many, and soon there were thousands—and all the ants continued to live in harmony with one another instead of staging the usual fights over territory that are quite normal between ants of the same species from different colonies. A super-colony emerged, in which there existed a kind of inviolable peace, but it continued to act with increasing aggression toward the outside world to feed the many hungry mouths within it.

Of course, this led to trouble with the long-established ants

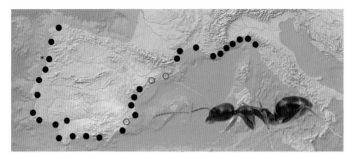

Two supercolonies of Argentine ants have spread along the coast from Portugal to Italy spanning 3,700 miles (6000 km). The map shows 33 spots where larger (black circles) and smaller (empty circles) groups of the super-colony have been found.

in the area. They were often significantly larger and stronger and had no intentions of abandoning their ancestral home to these newcomers. If it were simply a matter of brawn, the Argentine ants' conquest would have been doomed from the word go. But what the Argentines lacked in power, they made up for in speed and team spirit. They would be quicker than their rivals to root out sources of food and swiftly carry them off. The Argentines would have practically cleared the buffet table before any of the other species so much as got wind that there was something to eat. And where trouble arose, the threatened worker would call her extensive family for help, overwhelming even the scrappier ants with a flood of enemies.

The South was soon in Argentine hands.

IT'S THOSE GENES AGAIN

The conquerors' secret lies in their great number and unity. Instead of fighting among themselves, as is common for ants of the same species from different colonies, members of this invasive species act as one colony and turn their accumulated aggression outward. This is extremely effective but why does this behavior occur here and not in the invaders' country of origin?

Once again, it's all in the genes. To put it crudely, genes compete in the hope of multiplying. Not consciously, of course, but a moderately valuable gene that multiplies successfully will still be present generations later, while the most brilliant gene will be lost if it doesn't ensure it is passed on to the next generation. From the genes' perspective, the bodies of animals, plants, bacteria, and even people are little more than tools, helping them to reach their goal. Consequently, in a naturally developing animal society with a variety of different gene variants, there is a marked struggle between genes and their tools for the resources required for multiplication, from food to nesting sites to suitable partners. Each gene variant wants the best and the most for itself. And this causes stress with others from the same species that are, however, a little different.

It is precisely this slight difference that invasive species are lacking. In these groups, all individuals can be traced back to the animal or animals that washed up in their new environment. They all come from the same queen and therefore share the same genes. As such, there are no grounds for competition, separate territories, or closed-off nests. Even in a colony of billions of individual insects, all the members of the family are closely related, largely resembling one another genetically. The ants recognize this because all their sisters smell the same as they do. A worker may be from the Portuguese Atlantic coast, from Marseille, France, or the Italian Riviera, but its scent will remain the scent of its common ancestor. The scent of its supercolony.

Consequently, then, evolution had no need to reconstruct the genes of the Argentine ant and other invasive species in order to change their behavior. All it took was for the founding population of the new colony to be composed solely of a few insects with an extremely limited selection of gene variants. It was what is known as a genetic bottleneck, through which the species traveled as it transmigrated, making the offspring all alike.

For Argentine ants, this meant not only that the European supercolony remained one single family, it also meant that its offshoots in California, South Africa, Southern Australia, and Southern Japan remain so closely related that they view each other as sisters. When individual insects are exchanged, the ants are still able to cooperate with one another and share their food. With a little unintentional help from humans, the Argentine ant has succeeded in creating a mega-colony that spans the globe.

There is one exception, of course. Oddly enough, in Catalonia, the province where the people also want to split off from Spain, a second, smaller supercolony has established itself and competes with the rest of the European Argentine ants, fighting them for territory. At some point, there must have been a mutation in the gene that determines nest scent, and the Catalans were unable to resist the scent of independence.

Once an invasive species has built an empire, its neighbors are in for a rough ride. Regulating elements such as pathogens, predators, and tenacious competitors are lacking; there is nothing to keep these new kids on the block under control. And so, the invaders can run amok through the area, thoroughly unchecked. The Argentine ant not only decimates most other ant species in its vicinity, its accompanying aphids and scale insects cause great damage to local agriculture, bleeding crops dry. Without the flowers associated with these crops, pollinating insects cannot find food, or may be slain by the Argentines. The entire ecosystem goes haywire.

Things are particularly bad on Christmas Island.

SHAKING UP THE ECOSYSTEM

An external territory of Australia, Christmas Island is situated in the Indian Ocean and covered in great swaths of tropical rainforest. But large areas of this forest are more reminiscent of botanic gardens than unspoiled jungle, since the ground is so bare. This is due to the red Christmas Island crabs that live here in their millions and feed off low-hanging leaves and burgeoning seedlings. Once a year, the crabs migrate to the sea to lay their eggs. The journey of these red crustaceans is one of the most impressive natural sights that our world has to offer. At least, for now.

These creatures are threatened by *Anoplolepis gracilipes*, the yellow crazy ant. Yes, it's really called that. It takes its name from its habit of running this way and that when disturbed and its strikingly long, thin legs and antennae that make it look a little crazy. We don't know where the crazy ant originally came from; perhaps it came from China, India, or West Africa. There are colonies in Australia as well as on many islands in the Indian and Pacific Oceans, from Hawaii to the Galápagos Islands. At first, the ants integrated well into the ecosystem on Christmas Island. They continued this way for sixty years, as if they had always belonged there. But in the 1990s, something went badly wrong. Perhaps it was El Niño, whose altered ocean currents brought about a phase of extreme dry weather. The scarcity of water increased the concentration of nitrogen-rich nutrients in the trees' sap, which acted as a superstrength food for the scale insects that feed on them and whose honeydew is a major component of the crazy ants' food supply. So far, we have been unable to prove that it occurred in this way; what we know for sure is that the crazy ants have undergone an explosion in numbers since then. In some places, they exist at such a density that they form a carpet of ants, completely covering the ground beneath them. And that's just the ants that are out and about, the ants we are able to see. Nobody knows the real number of crazy ants on the island.

But their presence is hard to overlook. These masses of ants eat everything in their path, anything that cannot run away. Alongside insects, red crabs fall victim to the ants in great numbers. It is estimated that they have killed ten to twenty million of them. They spray formic acid in the crabs' eyes and on their joints, blinding and incapacitating them and causing them to die from dehydration or exhaustion, before tearing them to pieces and carrying the parts back to their nests. Even crabs whose territory lies in areas free of ants fall victim to them when they are forced to travel through the crazy ants' territory on their journey to the sea. Crustaceans, small mammals, and young flightless birds all fall prey to the ant hordes. Australian scientists are terrified that the ants could completely wipe out the Abbott's booby, which is exclusively endemic to Christmas Island.

Vast swaths of the once bright jungle are now unrecognizable. Since the ants decimate the crabs on a huge scale, there is no one left to keep the undergrowth down. As a direct consequence, seedlings grow tall, particularly those that previously would not have succeeded in encroaching among the trees. Plants in the nettle family spread, as does the giant African land snail, whose numbers the crabs had previously

The invasive yellow crazy ant has killed many of the red land crabs on Christmas Island, which once kept the undergrowth under control. The result is this sprawling jungle.

reduced. Not to forget the viscous honeydew, the quantity of which is now so great that the ants are no longer able to harvest it. Instead, it drips onto the plants below, sticking to their leaves and making them susceptible to mold and mildew, such that even larger trees ultimately die. Together, these two super-colonies of crazy ants and their accompanying scale insects are responsible for the desertification of 5 percent of Christmas Island in total. And there is no end in sight.

But it's not just crabs and trees that suffer as a result of these adventive ants. Nobody is safe from invasive species. Not even the king of the jungle.

THROWN TO THE LIONS

Many invasive ant species prefer warmer temperatures. They generally do not tolerate cold nights and frosty winters well, and so have only been able to spread in certain locations in Central Europe, such as in palm houses.

When I was working as a professor at the Ludwig Maximilian University of Munich, my colleagues and I observed that some specimens of the big-headed ant *Pheidole megacephala*— one of the one hundred worst invasive animals—had nested in the zoological garden there. The intruders considered this heated environment their territory and were constantly clearing away the other animals' food with the meticulousness for which ants are so well known. Whole highways of little black foragers closed in on feed meant for birds and tortoises. They had no respect, even for food left out for the lions, whose enclosure was in the palm house. If the keepers set the lions' meat down too early, before the lions arrived in the enclosure, it would soon be swarming with ants, and the kings of the jungle would turn their noses up at the meaty morsels and no longer want to eat them.

The ants were of great scientific interest to us, however, thanks to their excellent capacity for efficient recruitment.

At the time, we were pondering the question of how invasive ants were able to lead fellow members of the nest toward food sources so quickly. At our request, the director of the zoo was quite happy to permit us to set off on the hunt for these six-legged pests—a convenient opportunity to observe the ants without the need for a great expedition. We armed ourselves with buckets and spades and plunged into the artificial habitat, eyed mistrustfully by keepers and gardeners fearing for their carefully nurtured gardens. With great care and thoroughness, we dug around looking for the nests of the big-headed ants. We succeeded in finding a few nesting chambers and plenty of workers—but no queens. These would surely reside in deep chambers, and we would have had to dismantle half the palm house to bring them up and haul them off to the laboratory. Faced with this choice, the friendly zoo director concluded that the ants were the lesser of two evils, and we were obliged to leave without the queen.

Elsewhere, however, invaders are not limited to artificially heated spaces, capturing large outdoor sites that humans have built for their own use.

GOLFERS BEWARE: THERE'S A NEW KID IN TOWN

Americans love their golf courses. Large areas with neatly shorn grass epitomize order and leisure in this land of unlimited but carefully mown opportunity. The problem is, the red imported fire ant *Solenopsis invicta* likes these kinds of green spaces, too. And when a golfer approaches the nest, these little firecrackers know exactly how to make their displeasure known—unmistakably so, in fact, with unpleasant stings. In the year 2000 alone, 30,000 golfers in the southern states were hospitalized with ant bites, with over 100 deaths recorded, most of these from anaphylactic shock.

But fire ants have no business on golf courses, nor in fields or on arable land, nor in the gardens or parks of North

America. Like all invasive species, they have spread here because they are small, aggressive, and great in number, with a spirit of family cohesion that goes back to their genes.

Yet the cracks have already started to show in this big happy family. In the Gp-9 gene, the ants' genomes occur as two different variants, known as alleles, which are referred to using a capital B or lowercase b. Since every female ant carries one copy from her mother and one from her father inside her cells, there are queens and workers with the genetic configuration BB, Bb, and bb. This small difference has no effect on appearance, but it seems to influence an ant's viability, because insects with bb often die while still in the egg. The genetic variant b will therefore have difficulty staying alive. But what it lacks in fitness it makes up for with a good portion of selfishness.

Gp-9 also influences its carrier's scent, and when it comes to smells, ants are known to be picky, making short work of anything they find unpleasant. In a colony with Bb queens, the b variant causes the Bb workers to bite the young BB queens to death as soon as they hatch, wiping out the genetic competition. BB workers, who do not produce their own offspring, are spared and raise the Bb offspring. The evolutionary researcher Richard Dawkins, who popularized the concept of the selfish gene, calls genes like b, whose carriers recognize and support one another, "green beard genes." He chose this unusual name because these cliquey genes behave as if they have given their carriers a green beard; all green-beards make sure that fellow green-beards enjoy advantages while making life difficult for those with different-colored beards. Genes can clearly be pretty lousy to each other and they settle their petty squabbles through the bodies of the organisms which carry them.

They also reveal a whole lot about the history of a species. Historians are generally not all that interested in little creatures like ants, and when a species is so meddlesome that it becomes hard to ignore, its beginnings often lie so far back in the past that it's almost impossible to discover when and from where the insects migrated. In such cases, it's often helpful to look at the genes. DNA is a highly stable molecule, with repair mechanisms in the cell nucleus fixing almost all flaws or changes that arise—but only almost all of them. Occasionally, some deviation from the norm is preserved and, over time, these mutations accumulate within a population. The larger the genetic variety of ants in a region, the longer they have evolved there. A founding population selected at random, on the other hand, begins life in its new homeland with almost nothing. Analysis of the genes of fire ants across the world has now revealed that these insects demonstrate the greatest genetic variety in South America, which is probably where they originated. The incidences of them in the US, Australia, China, and the Philippines are, by contrast, genetically extremely similar. We can conclude, therefore, that they were transported there by humans—like almost all invasive species.

But it doesn't always have to be this way.

COLD-HARDY AND PROPRIETORIAL

The ant *Formica fuscocinerea* is one of the few examples of these little conquistadors to have exploited new areas without human intervention. At least, that's what they've done in Southern Germany. On their triumphal procession northward from the Austrian Alps, the ants have encroached on the Munich area. Always on the search for sandy regions, they have snatched control of the gravel banks of rivers such as the Isar, commandeering railroad embankments, sparsely vegetated roadsides and—much to the locals' horror— playgrounds. As many as 212 nest entrances have been counted in an area little bigger than a square yard, and all the inhabitants, from Dachau, Germany, to their home in Austria, belong to a single supercolony. Unlike the (at most) 250 invasive species of ants, *Formica fuscocinerea* do not mind cold winters. They are cold-hardy and will be sure to make their presence known the following spring. After all, plenty of large groups of these little conquistadors are only spotted when it is too late.

Which is how the invasive garden ant *Lasius neglectus* has been able continue its furtive spread across the world for so long, because it is practically identical to the common black garden ant, *Lasius niger.* The safest way to tell them apart is by the number of workers you encounter: if there are just a few insects scurrying along a trail, they are black garden ants. If, however, the trail is so crammed with foragers that it looks less like a beaten track and more like a multilane highway, you are dealing with the invasive garden ant.

Entomologists only discovered they were dealing with two different species in 1990, in an arboretum in Budapest. Presumably originating in the Black Sea area, the invasive garden ant was sent to parks and botanic gardens all over Europe in the soil accompanying plant cuttings, establishing offshoots in more and more towns and cities as time went on. They also outpaced the local species not only in numbers but also with their greed. Species of the genus *Lasius* commonly take possession of a food source, defending it against competitors who may later emerge, but if they find someone has already made a claim on it, they accept the other ant's right to the food. Not so with the invasive garden ant. In their view, all food is the property of their colony, so they will often fight over food which, strictly speaking, already belongs to somebody else. In doing so, they have expelled seventeen other species of ants from Budapest alone. And when they're done making themselves unpopular outdoors, they will happily appear in our homes, where they delight in nesting close to electrical devices.

But apartments, businesses, and hospitals are the preserve of another species of ant.

HEAT-LOVING GERM PROPAGATORS

The pharaoh ant *Monomorium pharaonis* is one of the smallest ants in the US and Europe, and certainly the most dangerous. Not because pharaohs are particularly big biters or stingers, but because their puny stature means they can fit through the narrowest cracks and hang out in the grimiest of places, cheerfully spreading all kinds of pathogens.

This species made its way around the world as a classic vagabond from the tropics, so it does not tolerate the harsh northern climate well and prefers to seek out cozy spots inside human homes. Its favorite spots are inaccessible cracks next to furnaces, hot-water heaters, or ventilation units (ideally in bakeries), or commercial kitchens or hospitals, where the table is always laid for dinner. Pharaoh ants are not picky eaters, gobbling up anything they find, from dead silverfish to sugar water to raw liver. Unfortunately, they also eat urine, feces, vomit, blood, and pus; the list of bacteria they pick up might remind medics vividly of their microbiology lectures. Only meticulous levels of cleanliness can keep these insects from accessing their favorite foods. Even well-sealed packets and closed refrigerator doors will not hold them back. And if hospital patients are unable to defend themselves, the ants have no qualms about crawling under dressings or casts and nibbling at the patients' wounds. They can infect the weakest, such as premature babies, and gravely ill or dying patients, with a host of unpleasant pathogens.

Pharaoh ants can plague our homes, too, and are a pest especially in Florida. They creep along gaps between hot water pipes, using them to get around the house. If one of your neighbors has pharaohs in their home, they will soon see them marching through the kitchen, hard at work, spreading salmonella. For this reason, any food which has played host to a little visit from the ants is better off in the garbage outside. And if your kitchen starts smelling like spoiled meat, the source might be the ants' personal supply store. They create these little depots when they are unable to carry everything back in one go. But if there is a constant abundance of food, they will forget about their stores and the crumbs of sausage and meat will begin to spoil.

Then—at the very latest—it's time to fight back against this invasion of little overlords.

TACKLING THEM WITH TECHNOLOGY

Unfortunately, this is where the problems really start. Despite all scientific wisdom and high tech, it's almost impossible to completely get rid of an invasive species. We have mostly abandoned the old methods of forcing intruders out by introducing a natural enemy from their country of origin—and with good reason. Numerous cases have shown how horribly wrong things can go when you intervene in an ecosystem without sufficient preliminary studies. A prime example are the Australians who, in the 1970s, hoped to introduce cane toads to tackle a plague of beetles in their sugar plantations in a biologically friendly way. A positive outcome would have been a reduction in the initial infestation of beetles. Unfortunately, all that resulted was an influx of poisonous amphibians.

Nevertheless, tests are currently underway to see whether yellow crazy ants could be managed using a newly introduced species. They are hoping to attack not the ants themselves, however, but their main source of food, the scale insect. A tiny wasp that shares the scale insects' Southeast Asian home lays its eggs exclusively inside this species and could decimate the population, bringing it down to an acceptable level. Only on little Christmas Island, of course—there are plenty of other species of scale insect living on the Australian mainland, meaning the ants could move on to a different species of livestock and continue growing their super-colony.

Another method of stopping these invaders in their tracks relies on a refined variety of chemical mace. Pest controllers in California are hoping to use artificial pheromones that mostly match the colony's scent profile but differ enough that a worker views one of her sister's sprayed with the scent as a foreign intruder, provoking civil war among the Argentine ants. In Japan, however, pheromones are employed to cause confusion by imitating the ants' trail markings and disrupting their navigation systems. Of course, traditional poisoned bait has also had a certain degree of success. But the effects never last long for any of these methods. It's impossible to kill all colonies entirely with this treatment. As long as a single queen survives, she can begin to rebuild her empire once the pheromones wear off, or the poison breaks down. And then the drama begins again.

But perhaps nature knows best how to fix itself.

A SURPRISE ENDING

In 2018, scientists in New Zealand reported that populations of Argentine ants in forty per cent of all areas examined had collapsed and, in many other regions, they had shrunk to just a few nests—just like that. Researchers made similar observations in the Seychelles, where the yellow crazy ant had completely disappeared in some areas, to the benefit of the indigenous species that immediately began to repopulate their lost territory. In Texas, too, some areas began to recover from the fire ants. Colleagues who had been following the development of ant communities in a natural park told me that there had been a good seventy-four different species in the park before the fire ants invaded, with only twenty-one remaining afterward. Meanwhile, however, the density of the fire ants in this area has dropped and the number of different species has risen again to fifty-three. We don't yet know the reasons behind this seemingly magical disappearance. But we have two ideas about the developments that may have played a part in it.

First, there are the external factors. Invasive ants are successful because they don't get sick and they don't get preyed upon. Both factors can change, however, if pathogens and insectivores have time to adjust to the newcomers. If, for example, a fungus discovers a taste for members of a super-colony, it can spread rapidly. All workers are practically identical and have the same defense mechanisms. If a pathogen can bring down one ant, the others have nothing new with which to tackle it. The recipe for success that is genetic similarity turns out to be a recipe for rapid mass mortality, as we know from our own monocultures.

The internal factors also draw on genetic monotonicity. As we saw in chapter 10, On Parasites and Slave-Makers, sometimes a species will split according to minute differences, producing a parasitic sister species that prospers at the cost of its sister species, exploiting it thoroughly. These social parasites can thrive wonderfully even in super-colonies. Species consisting solely of queens and males, and which contribute nothing to the community due to the dearth of workers, can weaken billion-ant colonies to such an extent that the ants revert to their original behaviors.

Like human empires, the empires of the ants are never stable for long. If only they could take turns.

BATTLE OF THE SUPERCOLONIES

The most spectacular way for an empire to fall is surely at the hands of another superpower. Alexander the Great swallowed the superpower of Persia, the Byzantine Empire fell to the Ottomans, and the Mongols conquered China.

A similar power struggle is unfolding in the warm southern states of the US. The supercolony of red imported fire ants seemed to have the area firmly in its grasp, but then *Nylanderia fulva*, or the tawny crazy ant, catches up with it again. The first sightings of this ant were made in 2002 in Houston,

Texas, and since then this species has spread across other areas of Texas, Florida, Mississippi, and Louisiana. Since neither the queens nor the males can fly, the course of their conquest has been a slow one. To compensate, the ants are much more meticulous when foraging for food than other species. They collect such a great quantity of food from an area that they could feed up to one hundred times as many ants as the competition over the same area.

Naturally, the local pests are no fans of these little upstarts. So the red imported fire ant follows the crazy ant wherever it goes, ready to fight—and loses. The *Nylanderia* has a secret weapon: an antidote that neutralizes the fire ant's venom. The secret? Plain old formic acid. If a fire ant sprays a crazy ant with its signature cocktail, the affected fighter stands on its back legs and pushes its abdomen forward so that it can reach its own glands with its mandibles. It takes a drop of formic acid and, using its forelegs, spreads this over its entire body. Once covered, it happily returns to the fray. This is a defense technique that myrmecologists have observed in both species' native home in South America and has been found to work even better in their adopted homes. The crazy ant is beating the fire ant back, yard by yard, state by state—much to the displeasure of the people living in these areas.

Unbelievably, people living in the invaded areas want the fire ants to come back. The ants view all terrain as their personal property but only get aggressive when you stray too close to their nests. The crazy ant, on the other hand, is not content with taking over yards, gardens, fields, meadows, wilderness, and golf courses—it wants to get into our homes. Once inside, it nests within the walls and intermediate floors, inhabiting air-conditioning units and household devices, exploring computers and televisions. On their travels through the home, the ants nibble at electrical cables, often causing them to short-circuit and are probably to blame for a number of house fires.

There is no defense yet against the crazy ant. They are not tempted by poisoned bait and they have no natural predators in the US. All people can do is make painstakingly sure that they have not brought any ants with them when traveling to *Nylanderia*-free areas from those afflicted with them. This wins them a little time until researchers find a way to beat them. Or until the next invasive species decides to take on the crazy ants' new empire.

Of course, this information is of little use to you if you have ants crawling all over your carpets at home.

BARRICADES AND AROMATHERAPY

The question people most frequently ask me when they discover that I study ants for a living is, "How do I get rid of the critters?" The answer is pretty simple: "It depends!"

If you're unlucky enough to have pharaoh ants nesting in your home, you should not waste time experimenting. You should call in a specialist who has experience with these pests. And you should opt to treat your whole home with something that lasts several months. Poison will not do it, because pharaoh ants avoid food sites where their sisters have had bad experiences. That's why professionals often set down poison containing a juvenile hormone, making the queens infertile. This will cause a supercolony to gradually age and die of collective infirmity.

But if it is just the usual ants tramping through your home from the garden or the street, I recommend stalking and spying on the workers to find out how they are getting into the house. They usually creep through tiny holes in windows or doors, or hairline cracks in brickwork. Simple fill these in with a suitable mixture from a hardware store and soon the stream of ants should dry up. Alternatively, or in addition to this, you could deter them with chemicals that stink to high heaven to ants but smell quite pleasant to humans. Herbs and oils such as basil, chili, grapefruit, coffee, lavender, orange, peppermint, rosemary, sage, tea tree, thyme, tomato, juniper, cinnamon, or lemon are good for this. Place, spray, or spread a little of these in their path, and with a little luck they will lose interest in picnicking between your four walls.

If you have ants living in your garden, you are in luck! They are free and tasty snacks for an eager army of pest-predators and soil improvers. Ants are also a perfect remedy for those times when boredom strikes. Simply take a magnifying glass and study the lives of your little neighbors. If you've read this far, you're now pretty well acquainted with ants, after all. And who knows, perhaps you'll be the first to discover a particularly bonkers behavior in a local species, as we'll turn to next.

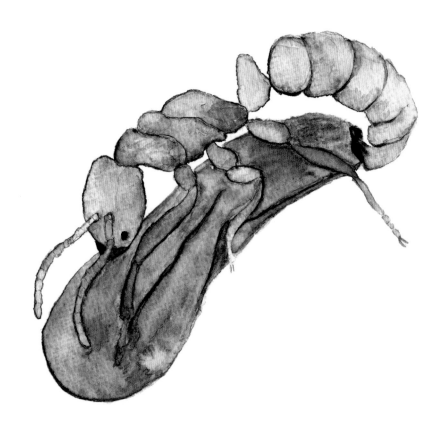

CHAPTER 13

CRAZY CRITTERS

A Hypoponera *male mates with a young queen still inside her cocoon.*

So far I've told you a lot about ants: queens with little say in their own realms; scouts who navigate by scent and polarized light; whole nations housed in hollow acorns or castles of leaves stitched together; ants who keep livestock or who are exploited by them; supercolonies who monopolize whole islands and conquer empires thousands of miles wide; nurses who heal the wounded with antibiotics; and much more. Some of this you may well have already known and some of it was probably new to you. But for ants, all of this is natural—and often normal to all of them.

But of course, there are also other stories: habits, traits, and idiosyncrasies observed in very few species —sometimes only in one—that are so bizarre, astonishing, or remarkable that I simply have to tell you about them. And few species grant us so close a glimpse into their unique inner lives as the four dots ant *Dolichoderus quadripunctatus,* which has four spots on its abdomen caused by a lack of pigmentation, allowing us to see its intestines without the need for a scalpel. See-through ants, who'd have thought it?

TAKING IT TOO FAR

If you want to take over the world, you've got to be able to adapt. This applies as much to the technology we humans use as it does to the natural adaptations that nature tests out on different species of ants. Depending on what's required, nature will choose a larger or smaller body, long or short legs, big heads or small heads, and use them to conquer environments as diverse as bare ground, the tops of trees, and the cases of electronic devices. Still, I sometimes wonder whether evolution has led some ants down a dead end.

The four dots ant lacks pigmentation in four areas on its abdomen, allowing you to see right inside.

Species of the African genus *Melissotarsus* are not quite of this world. Aside from the brief periods during which queens and males are outside, seeking a partner for copulation on the nuptial flight or founding a new colony, *Melissotarsus* prefers to keep to itself. It uses its extremely strong mandibles to gnaw pathways through the fresh wood of living trees, where it spends its whole life. It is one of the few ants capable of this. Most other species would find this wore down their mandibles; they would be unable to create a tunnel. They would have to wait until the wood was dry and chapped or moldy and rotten, or use tunnels nibbled out by beetles and their larvae. Most ants are content to make do with existing hollows underneath the bark. To them, wood is much like what caves were to early humans: welcome dwellings which you might fix up a little, but which you essentially took as they came.

Melissotarsus, on the other hand, is born a drilling machine. It's not just its masticatory organs, which are built for a creative existence inside wood. Its legs have also developed especially for this purpose. They are broad and flat, resembling the legs of bees. More striking, however, is the orientation of their central pair of legs, which stick up. Inside the nest, this allows them to crawl along the ceilings of corridors and anchor themselves as they gnaw their way through, like a tunnel-boring machine creating a new subway shaft. If they accidentally break through into the outside world, the hole is swiftly plugged with a stopper made of silk, which the adult workers produce from specialized glands on their heads. This is another of *Melissotarsus*'s specialties—other ants can only produce silk during a particular pupal stage, if at all.

These drilling ants are so brilliantly adapted to secret lives in their woody homes that a single tree can house colonies numbering an estimated 1.5 million insects beneath its bark. Still, their specially equipped lifestyle comes at a price. If a *Melissotarsus* ant finds its way out of the tunnel into the world outside the flat areas of bark, sand, or leaves, it is doomed. Due to its twisted legs, it can't walk outside its usual corridors; it can't even stand up. The moment it raises itself up on its four legs better suited to the outside world, it loses its balance and tips over. This makes *Melissotarsus* the only ant that can't crawl. Of course, you might wonder what these ants eat, since they are unable to go outside to forage for food. *Melissotarsus* ants live off scale insects, which lurk in their tunnel system in the wood, tapping the tree for sap. *Melissotarsus* lead isolated lives in their own little worlds—in stark contrast to some ants who are keen to conquer the very air above them.

SKYDIVING WITHOUT A PARACHUTE

Many scientists make the move to myrmecology out of sheer bewilderment. University of Louisville researcher Steve

Yanoviak was researching the effects of deforestation on the spread of mosquito-borne diseases. He regularly climbed up into the canopy of the tallest trees to collect mosquito larvae and hungry female mosquitoes, which he would ensnare by tempting them with a free meal of his own blood. Of course, this meant that from time to time he would enter the territory of tree-dwelling ants, who didn't think much of his research.

Fixated as he was by the mosquitoes, Steve would brush the attacking workers aside, causing them to fall from the branch and drop 33 yards (30 m) down. But one day, as he watched a sweep of ants falling in this way, he could hardly believe his eyes. Instead of falling straight down, the ants fell in a curve, landing a few yards below on the trunk of the tree. Steve was so amazed that he forgot the mosquitoes and spontaneously embarked on several more flight tests with the ants. Contrary to what you might expect, they did not tumble to the ground in an uncontrolled manner; instead, they were clearly able to guide their fall, steering right toward the tree they had just fallen from. Suddenly, Steve had a new field of research.

On further expeditions, he and other scientists discovered that these gliding ants were the species *Cephalotes atratus*, which we encountered in chapter 11, Physician, Heal Thyself, where we learned that nematodes turn them into red-berry-lookalikes. These ants do not use their gliding skills only when brushed aside by irritable researchers, they also use them to escape from powerful predators, such as birds. They fall intentionally and on purpose. At first, they fall straight down. The ants clearly spend this phase looking around because they soon orient themselves to the trunk of the tree and hurtle forward with their abdomens, their backs facing up, allowing for a controlled descent toward the tree. They can achieve 180-degree turns when necessary. How they manage to slow their rapid flight and cling to the bark as soon as they reach the trunk remains a mystery. In any case, the landing takes place at a furious speed and does not always succeed at first. Sometimes the little pilot bounces off and must make a renewed attempt a few yards down. Eventually, however, it grabs hold and crawls back up without so much as a breather. It takes less than ten minutes on average for a worker (marked with a dab of paint) to find its way back to where it started.

By no means do all ant species possess this ability to glide, but the skill is more widespread than researchers once thought. It doesn't require a particularly aerodynamic body type, occurring both in "flat" ants such as *Cephalotes* and more spindle-shaped species. It appears to depend more on living conditions. Ants inhabiting the treetops exhibit particular skill as pilots. Good for them. When a worker does fall to the forest floor, it's usually all over for her. Since all trees look a little alike and she has no scent trail to latch onto, she has no way of orienting herself and will never find her way back to the colony. It's even more risky when such a drop ends on the floor of a forest that is flooded for much of the year. An ant that finds itself in water after a fall is guaranteed to be fish food. So, it's worth the workers' while to get their pilot's license. Even if they don't have wings.

. . . ONE GIANT LEAP FOR AN ANT

Letting yourself fall is a pretty easy method of taking to the air, but the trap-jaw ant's favored method is much more energetic. These ants have extremely strong mandibles, which they carry in front of them, wide open. This not only looks terrifying, the jaws themselves function like loaded weapons. A powerful sphincter muscle inside the ant's head is held tensed, unable to release the energy stored inside it because the mandibles are locked. Once a sensory hair on the mandibles is stimulated, this sends a signal that unlocks the ant's jaws, causing the forceps to snap together like a mousetrap. Actually, let me correct myself: They snap together much, much faster than that. The mandibles of *Odontomachus bauri* snap together at a speed of 70 yards (64 m) per second—that's 143 miles (230 km) per hour, as fast as a sports car on an open highway. They snap together in 130 millionths of a second—a bullet would only cover 4 inches (10 cm) in that time —releasing a force equivalent to 300 times the ant's bodyweight, akin to a human lifting five elephants.

It's an impressive physical performance and an interesting numbers game, but the ant has other things in mind when performing this trick. It usually uses its snap trap to catch prey, of course. Trap-jaw ants hunt alone, pursuing nimble prey such as springtails. They slowly creep up on their victim until they are close enough to grab them. Every failed attempt means the hunter must start again from square one, wasting precious time. Yet her turbo bite almost never fails, often slaying her prey in one fell swoop. Quite simply, there's no creature fast enough to flee the danger zone in time. The ant's snapping jaws are one of the fastest phenomena recorded in the animal world.

Sometimes the hunters become the hunted, so the ants also use their tensed jaws for self-defense. Little aggressors such as spiders are often flung backward by the force of their mandibles, while larger assailants such as myrmecologists marvel at the aggressive, flying warriors that they suddenly see flocking toward them. As it turns out, these snapping jaws are also excellent at catapulting the ants through the air. All the ant has to do is point its jaws at the ground and bite down. These little fighters will fly over 8 inches (20 cm) on average, but distances of up to 16 inches (40 cm) have been recorded. Their flight is not controlled in any way, however, with multiple ants tumbling through the air at once. Still, there's plenty of time for them to orient themselves.

The snapping jaws of the trap-jaw ant are lightning-fast and can trounce prey such as springtails. Alternatively, when danger strikes, the ants can use them to catapult themselves to safety.

This built-in slingshot has many benefits. So many, in fact, that different species of ants have invented similar solutions at least four times, independently of each other. The jump always works the same way, in principle, but some aspects differ. The genus *Myrmoteras*, for instance, uses its muscles to build up so much tension that the hard cuticle on its head becomes slightly deformed, acting as an additional spring. This is a little like an Olympic high jumper bending her knees to make it over the bar. It just goes to show—when ants do something, they do it right.

MARRIED OFF YOUNG

We all know couples who have been together since they were kids. But have you ever heard of kids being conceived before their mothers are even born? That's another thing that happens in the world of the ants.

Hypoponera opacior is a ponerine ant species distributed across North and South America. These tiny ants have a particularly fascinating sex life, as my work group and I observed in a population in Arizona. These unassuming creatures live in the soil under large rocks or boulders. Collecting ants of this kind means dragging huge boulders this way and that from morning till night. After several weeks of sweat and toil, our hands are usually covered with calluses. It helps to use a sturdy spade as a lever. A few years ago, we bought such a spade for a good price in Mexico, because the field station was next to the Mexican border. On the way back, our little shopping trip caused a stir with the US border officials, who spent a good while thinking over our answer to the question as to why we had bought a spade in Mexico ("To collect ants!"), thinking we were taking them for fools.

With or without a spade, once you've shifted a rock and exposed a colony, you have to lie flat on the ground to suck up the ants using an aspirator. To do this, you must carefully follow the ants' network of tunnels, which are often populated with several nesting chambers. I imagine archaeological digs are quite similar in meticulousness.

Here is where you have to get your timing right. Flying males and queens traditionally swarm for their nuptial flight in early summer, which is interesting, of course, but nothing new. It gets much more exciting in late summer, when a second generation of reproductive insects has hatched. This time, however, they do not have wings and must reproduce inside their home nest. The males creep out of their pupal shells earlier than the young queens, instinctively knowing that they will be fertile for only a brief time. So, they get to work right away, though their partners have not yet hatched—fertilizing the females which are still in their cocoons through a small opening at the end.

In this species, the pupae of young queens, males, and sterile workers look remarkably similar and are almost the same size. In their fervor, males often fail to mate with the pupa of a future queen and try instead to fertilize one of their sterile sisters—or one of their brothers. In the first scenario, nothing happens because workers do not have ovaries or sperm pouches, so are unable to reproduce. But for a male still residing in its cocoon, the process can be fatal. The misguided male appears to realize his mistake, because he doesn't really copulate with his brother, he simply grabs onto his pupa. But the affected pupa often dies, causing the sneaking suspicion that it's not a mistake after all but a deliberate act of fratricide, an attempt to see off the competition for the real young queens as soon as possible. This is supported by the fact that we mostly observed this behavior when there was only one male active in the nest and there were no other males around to see to the ladies. Fratricide allows the ant to spend longer as the cock of the walk, spreading his sperm among a larger number of future queens. This brotherly embrace kills eventually, but too slowly to really prevent any real competition. Soon, countless males are out and about in the nest.

The offending ants shift their strategy in response to their love rivals. Now it's a matter of mating with a pupa of the right sex. These are mounted and fertilized, with copulation lasting from as little as a few minutes to as much as two days. To be on the safe side, the exhausted father-to-be then sits outside his bride's pupa to make sure no other males make a donation after him. Amazingly, there are no aggressive scenes between the males during this procedure. Whoever gets there wins the pupa fair and square, or so the rules seem to suggest. Trouble only emerges when the young queen has hatched, because she's not too happy about mating with a male. A young queen usually makes her way through the nest, dragging the offending male along behind her until he desists.

Three weeks later, the show finally comes to an end. All the new queens have hatched and coupled up and the soft-skinned males have served their purpose and are killed by the workers. The young queens live on, but often remain in their mother's nest. Sometimes they receive a few workers as a dowry and make their way through the local tunnels, off to build new sister colonies. Their flying sons and daughters will take to the air the following July and ensure that their species' unusual mating ritual continues farther afield.

A GENETIC REBELLION

Some marriages are short on domestic bliss. But I would bet good money that neither the human world nor the animal kingdom contain another couple who live as far away from each other as the queens and males of the species *Wasmannia auropunctata*, the electric ant or little fire ant.

These ants are an invasive species native to South America that have settled in North America, West Africa, and countless islands in the Pacific, Indian, and Atlantic Oceans, frequently terrorizing the locals. On the Galápagos Islands, they have multiplied to such an extent that they have come to pose a real danger to the giant tortoise, killing their young and infesting their eyes and anuses. But this isn't about that, nor is it about the astonishingly painful stings that these little ants can dispense; this is about the family life of the little fire ant.

It could be quite a cozy little setup, having several queens in each nest, the only ants in the colony able to lay eggs and produce offspring. The work is done by sterile females, who hatch out of fertilized eggs, while the males have little more to do in their short lives than deliver their seed to young queens.

The reproductive insects do a great job of this, but afterward they seem to endeavor to dupe the others. What initially appears normal may, on closer inspection, turn out to be quite bizarre.

Let's start with the queens. In other species, the young queens emerge from fertilized eggs that have been better fed during the larval stage than those of normal workers. However, queen fire ants boot the males out of their royal line altogether. Instead of hatching out of fertilized eggs containing genetic material from both a mother and a father, the young queens hatch out of unfertilized eggs containing a set of genes solely from their mothers. They are 100 percent clones of the queen, having no fathers at all. Like mother, like daughter, you might say. And she's got good reason to be pleased. As we've seen time and again in this book, nature is all about passing on as much of your genetic material as possible—and you don't get higher than 100 percent. With this trick, little fire ant queens have scaled the peak of genetic selfishness. If they were capable of pondering things like heredity, they would be pretty pleased with themselves.

If it weren't for their husbands, that is, who are sly customers when it comes to genetics. From what we've discussed so far, we'd assume that it would be the males who hatch out of unfertilized eggs, not the queens. But the little fire ant does everything topsy-turvy: In this species, a male hatches out of a fertilized egg. And when he does, he is the very model of his father—not only in looks, but also in his genes, because his mother's genetic material gets lost somehow during the egg's development. It's a battle of the sexes on a cellular level, with the father winning out in the eggs, which develop into males. All the mother is good for is packing the father's genes into nice, viable little packages.

So, we have got females with no fathers, genetically speaking, and males who do not bear a single one of their mother's genes. It might sound a little strange, but from a biological standpoint, it's actually all that is required to classify these reproductive animals as separate species. All that now connects the two are sterile workers who—quite ordinarily—carry the genetic material of both of their parents. And it's a good thing, too, because this mixture of genes creates the variation that promises better adaptability to different living conditions and changes, as well as better resistance to pathogens. These are particularly positive attributes for children who are essentially the colony's staff, doing all the work and having to survive the dangers of the world outside the nest.

However, just as biologists were beginning to worry about these creatures' future survival as one related species, positive news arrived from myrmecologists following the activities of the little fire ant out in the wild, in their native home. The queens and the males were certainly cooking up their own genetic soup in the wild, but they also got along much better, often mixing their genes in the usual way. Perhaps the unity of the species can still be saved. The island invaders may perish of their own accord, thanks to the eccentricities of the emigrants themselves. This romantic drama is one to watch.

KEEP OUT!

There are species of ant whose names I need only mention, and you'd probably be able to tell me the rest yourself. *Colobopsis truncata* (previously *Camponotus truncatus*—the kinship of ant species is a very dynamic field of research) is one example. Its common name is the cork-headed ant.

If you are now imagining an ant with a cork for a head, you're on the right track. This ant looks exactly as its name suggests. And if you ever find yourself in Southern Germany, with any luck you'll see one running along the branches of a tree. This species likes the heat, preferring to make its home in the Mediterranean, but *Colobopsis truncata* is also an unwelcome

Cork-headed ants use their heads to stopper the entrance to their nests.

visitor to Germany, carried over the Alps with fashionable Mediterranean plants, and has managed to settle in some regions without making itself unpopular. We regularly find it in the warmth-loving oak forests of the Rhine plain around Mainz, Germany. It is the only species of ant in Germany with a real subcaste of soldiers and it is only this group, alongside the queens, that are distinguishable by their stopper-shaped heads. (There is also a species of *Colobopsis* in the southeastern region of the United States, *C. impressa*, which looks just the same, so you can check oak forests there, too).

These small colonies of only a few hundred insects live in systems of tunnels that they fashion inside the dead branches or under the bark of trees. The colony is often split over several nests, each of which has only two or three narrow entrances. To keep unwanted guests from waltzing in, there is always a soldier on duty—larger openings will have two or three—who are able to seal the hole simply by shoving her custom-fit head into the opening. Her antennae lie laterally, pointing forward so that the tips are slightly visible. A worker will "ring" them when she returns from foraging, touching the tips repeatedly with her own antennae. If the password is correct, the door-keeper ant will make room for the worker to squeeze past. It's a very elegant way of protecting the nest, even if the doorkeepers themselves don't look all that elegant to us. If nothing else, they survive to see another day. And that's more than the gate-keepers of other species can hope for.

There is nothing the Brazilian ant *Forelius pusillus* fears more than a nighttime attack on its nest in the ground beneath a sandy sugarcane plantation. So, every evening it carefully locks itself away, hiding any trace of habitation from passers-by. Several workers bury the entrance with sand, fetching it grain by grain, before finally kicking more sand over it using their back legs, like dogs. The only problem is that the ants shut themselves out of the nest in the process.

This doesn't seem to matter to them, though. On the contrary, as soon as this handful of workers have finished their task, they run off into the distance, never to return. The presence of a dead ant could cause an attacker to suspect there might be more in the area. Instead, the ants who sealed the nest wander off to face their certain deaths, because, whether they starve or dry out, almost none of them will survive the night outside. Biologists describe this self-sacrificing behavior as proactive defense, and it's an unusual behavior even for these selfless ants.

But things are about to get much more spectacular.

KA-BOOM!

Sacrificing yourself to save the colony is an instinctive matter of honor for ants. Some of them take up John F. Kennedy's call to "Ask not what your country can do for you—ask what you can do for your country" in the most drastic way imaginable. The Malaysian ant *Camponotus saundersi*'s response is to blow itself up.

Now, even ants are not stupid enough to blow themselves up for fun or out of boredom. Indeed, it's their last resort in a desperate fight for survival. In principle, it's comparable to the self-destruct mechanism that starship captains deploy in science fiction films when their cruiser is captured. The ants' enemies might be birds or mammals who fancy a crunchy treat, but they are often weaver ants hoping to challenge the ants for their territory.

The ants' self-destruct mechanism is located in an unusually large gland in the mandibles, stretching the length of the ant's body, and which contains a sticky, toxic cocktail. If the ant feels threatened and unable to escape, it will

When the going gets tough for Camponotus saundersi, *it will blow itself up and use a toxic cocktail from one of its glands to stick to its attacker.*

tense its muscles so powerfully that its abdomen will burst open like a water balloon, tearing along the joints of its individual limbs, ripping open the gland in the mandibles, and spraying viscous poison in all directions. Larger predators lose their appetite at the disgusting smell and taste of the poison, which makes them more likely to leave this species to crawl along undisturbed in the future. Smaller attackers like weaver ants, however, find themselves coated in a sticky, corrosive substance, white to yellow-green in color. In emergencies, *Camponotus saundersi* will take its opponents with it to the grave.

Either that, or it will drive myrmecologists mad.

These insects are particularly prone to exploding when you're trying to collect them with tweezers. After hours of painstaking, exhausting work, you'll often have nothing more to show for it than sticky tweezers and dirty hands. Sometimes the most fascinating creatures are the hardest to capture.

SWARM INTELLIGENCE FOR ALL

We've mentioned science fiction and technology already. In fact, our high-tech civilization could probably benefit from the wisdom of these otherwise not overly brainy ants, especially when it comes to coordinating separate systems and working toward a common goal. It might sound a little abstract, but ants have solved some technological problems through several millions of years of trial and error.

In the past, it was easy: You'd have a programmer with a particular problem and he would know the constraints involved in great detail. Say he had to teach an automatic forklift truck for collect parts from preassigned places on shelves in the same order, time after time. He would have written a fixed program, the instructions for which the robot would have followed to the letter. It was quick, everyone was happy, and it worked, so long as the task never changed.

The technology of the future, however, faces far greater challenges. It's often not possible to know the conditions in the field of application in advance and changes can emerge at any time. Countless little units are used instead of large, individual robots. These have to work sensitively, as a collective, though each mini robot has memory capacity for only a handful of commands. These are the challenges facing engineers researching the tiny little robots that might one day patrol the surface of spaceships, repairing damage caused by micrometeorites, search the wreckage of homes in earthquake zones, or even travel through the body's bloodstream, fighting pathogens and cancer cells.

These scenarios are pipe dreams for the time being, but they are already contemporary science—where myrmecology, nanotechnology, and AI research meet to discover how ants perform such astonishing feats despite their simple brains, test these findings as flexible programs in theoretical models, and one day translate them into actual artificial swarms. Mechanical ants, able to independently perform intricate tasks.

Ant-think has already been implemented in some places. For instance, small programs automatically test the speed at which data is transmitted between interface nodes on the internet. These are scouts of sorts, traveling from one node to the other, measuring the time elapsing in between. If there is a lot happening on the line, the journey will take longer, and the program will furnish it with fewer "pheromones" in the overview table. If they get through quickly, the journey will receive a good "pheromone score." Actual data packages such as emails or websites, which are to be directed toward a destination, are then led along the paths with the strongest "pheromone trail." And since the situation is always changing, these values expire with time, much like real pheromones, so that once-poor pathways can later prove themselves and good ones are required to provide proof of their continuing quality.

Several telecommunications companies employ this method to test their networks—a method developed by ants to find the shortest path to a food source, when familiar turnings are often blocked and new ones emerge.

So don't be surprised if the ants in your garden keep finding their way into your kitchen, even when you thought you'd succeeded in filling in all the cracks and gaps in the brickwork. These little creatures are already working with the algorithms of the future.

AND, FINALLY, THE WEATHER

Despite their sophisticated technology, we should be careful about when to follow the ants' example and when to find better solutions, as their response to flooding after heavy rain will show.

Floods are no fun for anyone, including ants. They are not an infrequent occurrence either. What might seem to us like a couple of drops could equate to a flash flood in one of their underground nests, submerging corridors and chambers and often causing the nest to cave in; in worst-case scenarios, the brood or even the whole colony can drown. So, if you weren't smart enough to build your nest in a spot protected from the rain, there are a few ways you might try to protect it against floods.

Dams, for one. This idea is so obvious. It's not just us humans who use walls of earth to keep large bodies of water away from our homes—different species of distantly related South American ants do it, too. In the dry months, however, grazing livestock would destroy all the ants' hard work, so they do not build their dikes until the beginning of the summer rainy season. Within a matter of days, rings emerge out of the ground, measuring half an inch or more, and in the middle of these rings lie the dry entrances to the ants' nests. Other ants simply extend the channel from their nest exit, creating a kind of submarine turret that towers over the surface of the puddle. They use whatever building materials they can get: earth, mud, sand, and plant remnants. If the sun comes out between showers, it dries out the structure to create really solid defenses before the next rainfall. It is very similar to our own dams, dikes, and defensive walls.

It's not just humans who have hit upon the idea of using dams to hold back water. In South America, several species of ant use this method to protect the entrances to their nests.

By contrast, the methods employed by the Southeast Asian species *Tetraponera binghami* and *Cataulacus muticus* are less suited to imitation. Both species build their nests not on the ground but in the hollow stems of the giant bamboo. The external walls of these are waterproof, but during heavy tropical rainfall, rainwater can flood in through one of the entryways into the nest, just as it would through an overlooked kitchen window. The only thing to do is shut up shop at once, and the workers use their heads to stopper the opening in the hope of holding back the tide. Sometimes it works, and sometimes it doesn't.

If the water drips past the ants' little bodies and the nest is swamped, it is time for Plan B: The water has got to go. Since ants do not have buckets or pumps, they employ the only tool at their disposal: their own bodies. They set to drinking the water as slaking an incredible thirst, filling their crops to the brim with the unwanted moisture. Both species have developed their own methods for getting rid of it. *Tetraponera* crawls to the entrance of the nest like a living hose pipe, leans over the edge and spits the water back out. *Cataulacus*, however, seems to swallow the water and offload it outside the nest in a big, collective peeing session. Both varieties are particularly wonderful to observe when you spray a little colored water into their bamboo stem—the first ants will emerge a little later to spit or pee out the colorful water. Each worker manages to remove around 0.6 microliters of water. It takes more than fifty insects to bail out the quantity contained in an average tropical raindrop. It's no wonder it can take three days before the nest is dry again.

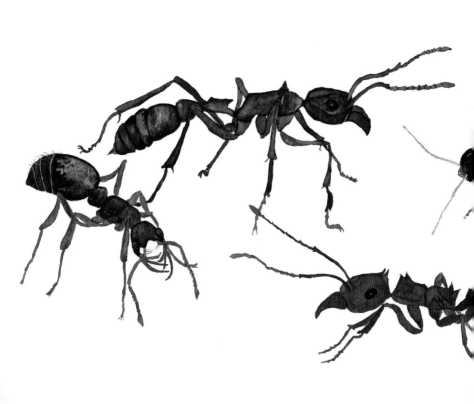

A
JOURNEY'S
END

*Not all ants are created equal. These are a few
that caught my eye in the Peruvian jungle.*

I am on a plane again. After several exhausting but illuminating days, it's time to head home. As the plane taxies along the runway, I find that I'm already wondering how I'm going to classify and evaluate my colleagues' and my observations. This time around, we were only in the rainforest to carry out field studies. Unlike other previous expeditions, we're returning from this one without any ants. It makes everything easier.

I smile when I think of the flight back from an earlier research trip where we had been hard at work collecting ants. Remember the aspirators we used to capture them? They're essentially just glass containers with two rubber tubes attached. Customs in Los Angeles decided ours looked suspiciously like a vaporizer for illicit drugs and we found ourselves being politely but quite firmly invited for further questioning at security control. Apparently, our answers weren't especially convincing. The officers thought it was all a ridiculous excuse: sucking up ants, removing their brains, and analyzing them for active genes? You would have believed us, wouldn't you? The US security officers, as it transpired, did not. We would have missed our plane had it not been for one of my colleagues who suddenly remembered that he still had a small plastic container filled with ants in alcohol in his trouser pocket. He swiftly drew out the piece of evidence and held it up to the officers. They took a good look at it and—thankfully—were finally convinced of our sincerity, and we were able to scurry off to our gate just in time.

We don't usually transport ants in our trouser pockets. But they don't require a mammoth effort akin to bringing along a rhinoceros as a carry-on. When transporting live ants, however, we must be considerate of our new pets' needs. Accommodation is relatively simple: they travel in a standard resealable freezer bag with a little leaf litter and some moist cotton balls. They have biscuits and a little ham as provisions for the journey. Heat poses the greatest threat to ants during transportation, so we stack the freezer bags inside cool bags, which are sometimes equipped with ice blocks. We aren't even allowed to leave them in the car to stop by the shops, especially in areas like Arizona. We take them with us wherever we go—to the supermarket, to visit friends, or to colleagues' labs. Just as many women are never to be seen without their handbags, I can be found tightly gripping my bags of ants. Except on planes, of course. That's why we book summer flights leaving in the evening, with no layovers, where possible, so that planes with ants stowed in the hold aren't kept waiting for hours in full sun. Thanks to these precautions, almost all our ants to date have made it safely across the Atlantic. None of our ant-laden luggage has ever been lost. Though we did have one close call.

On this one occasion, we had gathered on a site in Arizona boasting thousands of ant colonies for an extensive project carried out by one of my doctoral students. The five of us spent three weeks walking through the desert, cracking open countless rocks and using our aspirators to suck up colony after colony. Customs waved us through without any trouble, the plane was on time, and we arrived at the baggage carousel expectantly, exhausted but content, and ready to receive our quarry. Suitcase after suitcase slid along the carousel before finding its owner and making room for more luggage. We already had most of our belongings by now; all that was left was the bag containing the ants. The goods slid slowly out of the plane, the belt gradually clearing, but there was still no sign of the ants. Our earlier confidence gave way to nerves, then increasing panic. It simply couldn't be happening—after all that effort and stress, our ants were the only item of luggage on the whole plane to be lost? It couldn't be real, could it? But then the belt stopped moving. It was bare. And our ants had disappeared.

At this point, I have to say three cheers for the airline's friendly ground staff and their computer system. Faced with a small group of desperate myrmecologists, the woman at the desk was able to assure us that the bag containing our ants had definitely arrived in Frankfurt and would be waiting for us somewhere in the airport building. The question was where? It was time to employ years of training in careful and persistent tracking. We eventually found the missing bag in the corner of the Baggage Arrivals hall. Perhaps a fellow passenger had mistaken it for their own, taken one good look and been momentarily baffled by what had become of their socks and shirts. Either way, we had our ants back; my doctoral student snatched up his study objects and ran off with them, as if he planned to run all the way to Mainz, keeping them firmly in sight until he got there.

The seatbelt sign has been switched off. I carefully tip my seat back. A few more hours and we will be back in Germany. Tomorrow we'll begin evaluating our observations. And another task awaits: With my coauthor's help, I plan to write a book to teach people all about ants and why they are the most fascinating, impressive creatures on earth—for a myrmecologist like me, at least.

And now, I hope, you, too.

REFERENCES

To prevent the list of references from being thicker than the book itself, this is a summary list of references organized by chapter and specific research topic—a point of departure for readers who want to dive into the scientific literature on ants.

INTRODUCTION: SMALL BUT MIGHTY

WIDER READING ON ANTS

Bourke, A. F. G., & Franks, N. R. (1995). *Social Evolution in Ants.* Princeton University Press.

Fisher, B. L., & Cover, S. P. (2007). *Ants of North America: A Guide to the Genera.* University of California Press.

Gordon, D. (2010). *Ants at Work: How an Insect Society Is Organized.* Free Press.

Hölldobler, B., & Wilson, E. O. (1990). *The Ants.* Belknap Press.

———. (1994). *Journey to the Ants: A Story of Scientific Exploration.* Belknap Press.

———. (2008). *The Superorganism: The Beauty, Elegance, and Strangeness of Insect Societies.* W. W. Norton & Company.

Keller, L., & Gordon, E. (2009). *The Lives of Ants.* Oxford University Press.

Moffett, M. W. (2011). *Adventures Among Ants.* University of California Press.

Seifert, B. (2018). *The Ants of Central and North Europe.* Lutra.

THERE ARE APPROXIMATELY TEN QUADRILLION ANTS WORLDWIDE

Fittkau, E. J., & Klinge, H. (1973). On biomass and trophic structure of the central Amazonian rain forest ecosystem. *Biotropica, 5,* 2–14.

Rice, E. S. et al. (2017). *Dr. Eleanor's Book of Common Ants.* University of Chicago Press.

Wilson, E. O., & Hölldobler, B. (2005). The rise of the ants: A phylogenetic and ecological explanation. *Proceedings of the National Academy of Sciences USA, 102,* 7411–14.

ANTS HAVE EXISTED FOR AT LEAST 100 MILLION YEARS

Barden, P., & Grimaldi, D. A. (2016). Adaptive radiation in socially advanced stem-group ants from the Cretaceous. *Current Biology, 26,* 515–521.

Rabeling, C. et al. (2008). Newly discovered sister lineage sheds light on early ant evolution. *Proceedings of the National Academy of Sciences USA, 105,* 14913–17.

EARTH CAN COPE WITHOUT HUMANS

Holmes, B. (2006). Earth without humans. *New Scientist, 192,* 36–41.

Weisman, A. (2007). *The World Without Us.* Thomas Dunne Books.

ANTS ARE IMPORTANT FOR THE ECOSYSTEM

Del Toro, I. et al. (2012). The little things that run the world revisited: a review of ant-mediated ecosystem services and disservices (Hymenoptera: Formicidae). *Myrmecological News, 17,* 133–46.

Lach, L. et al. (Eds.) (2010). *Ant Ecology.* Oxford University Press.

Parr, C. L. et al. (2016). Suppression of savanna ants alters invertebrate composition and influences key ecosystem processes. *Ecology, 97,* 1611–17.

Philpott, S. M., & Armbrecht, I. (2006). Biodiversity in tropical agroforests and the ecological role of ants and ant diversity in predatory function. *Ecological Entomology, 31,* 369–77.

Way, M. J., & Khoo, K. C. (1992). Role of ants in pest management. *Annual Review of Entomology, 37,* 479–503.

CHAPTER 1: TAKE ME TO YOUR LEADER!

THERE ARE OVER 16,000 SPECIES OF ANTS WORLDWIDE, MANY OF WHICH HAVE YET TO BE RECORDED

Larsen, B. B. et al. (2017). Inordinate fondness multiplied and redistributed: The number of species on earth and the new pie of life. *Quarterly Review of Biology, 92,* 229–65.

Lau, M. K. et al. (2019). Draft *Aphaenogaster* genomes expand our view of ant genome size variation across climate gradients. *PeerJ, 7,* e6447.

Mora, C. et al. (2011). How many species are there on earth and in the ocean? *PLoS Biology, 9,* e1001127.

Wilson, E. O. (2007). *The Insect Societies.* Belknap Press.

ANT COLONIES ARE DIVIDED INTO CASTES

Rajakumar, R. et al. (2018). Social regulation of a rudimentary organ generates complex worker-caste systems in ants. *Nature, 562,* 574–77.

Weitekamp, C. A. et al. (2017). Genetics and evolution of social behavior in insects. *Annual Review of Genetics, 51,* 219–39.

Wheeler, D.E. (1991). The developmental basis of worker caste polymorphism in ants. *American Naturalist, 138*, 1218–38.

Wilson, E. O. (1978). *Caste and Ecology in Social Insects.* Princeton University Press.

THERE ARE MANY WAYS BY WHICH THE SEX OF AN ANIMAL CAN BE DETERMINED

Bachtrog, D. et al. (2014). Sex determination: Why so many ways of doing it? *PLoS Biology, 12*, e1001899.

Beukeboon, L. W., & Perrin, N. (2014). *The Evolution of Sex Determination.* Oxford University Press.

Shapiro, D. Y. (1990). Sex-changing fish as a manipulable system for the study of the determination, differentiation, and stability of sex in vertebrates. *Journal of Experimental Zoology, 256*, 132–36.

Viets, B. E. et al. (1994). Sex-determining mechanisms in squamate reptiles. *Journal of Experimental Zoology, 270*, 45–56.

Weeks, A. R. et al. (2001). A mite species that consists entirely of haploid females. *Science, 292*, 2479–82.

FEMALE ANTS HATCH FROM FERTILIZED EGGS, MALES FROM UNFERTILIZED EGGS

Kureck, I. et al. (2013). Similar performance of diploid and haploid males in an ant species without inbreeding avoidance. *Ethology, 119*, 360–67.

Miyakawa, M. O. et al. (2018). The doublesex gene integrates multi-locus complementary sex determination signals in the Japanese ant, *Vollenhovia emeryi. Insect Biochemistry and Molecular Biology, 94*, 42–49.

Queller, D. C. (2003). Theory of genomic imprinting conflict in social insects. *BMC Evolutionary Biology, 3*, 15.

Trivers, R. L., & Hare, H. (1976). Haploidiploidy and the evolution of the social insect. *Science, 191*, 249–63.

QUEEN ANTS STORE SPERM IN A SPERM POUCH

Baer, B. et al. (2006). Sperm storage induces an immunity cost in ants. *Nature, 44*, 872–75.

Chérasse, S., & Aron, S. (2016). Impact of immune activation on stored sperm viability in ant queens. *Proceedings of the Royal Society B, 285.*

den Boer, S. P. A. et al. (2009). Prudent sperm use by leaf-cutter ant queens. *Proceedings of the Royal Society B, 276*, 3945–53.

Gotoh, A. et al. (2009). Evolution of specialized spermatheca morphology in ant queens: Insight from comparative developmental biology between ants and polistine wasps. *Arthropod Structure & Development, 38*, 521–25.

QUEEN ANTS CAN LIVE UP TO 30 YEARS OF AGE

Keller, L., & Genoud, M. (1997). Extraordinary lifespans in ants: A test of evolutionary theories of ageing. *Nature, 389*, 958–60.

Kramer, B. H., & Schaible, R. (2013). Colony size explains the lifespan differences between queens and workers in eusocial Hymenoptera. *Biological Journal of the Linnean Society, 109*, 710–24.

Tschinkel, W. R. (2017). Lifespan, age, size-specific mortality and dispersion of colonies of the Florida harvester ant, *Pogonomyrmex badius. Insectes Sociaux, 64*, 285–96.

ANTS DEVELOP FROM EGGS INTO LARVAE, THEN INTO PUPAE BEFORE EMERGING AS ADULT ANTS

Schultner, E. et al. (2017). The role of brood in eusocial Hymenoptera. *Quarterly Review of Biology, 92* (1), 39–78.

Verza, S. S. et al. (2017). Oviposition, life cycle, and longevity of the leaf-cutting ant *Acromyrmex rugosus rugosus. Insects, 8*, 80.

YOUNG WORKERS TAKE CARE OF THE BROOD, OLDER WORKERS PERFORM RISKIER WORK OUTSIDE THE NEST

Besher, S. N., & Fewell, J. H. (2001). Models of division of labor in social insects. *Annual Review of Entomology, 46*, 413–40.

Hölldobler, B., & Wilson, E. O. (1986). Ecology and behavior of the primitive cryptobiotic ant *Prionopelta amabilis. Insectes Sociaux, 33*, 45–58.

Kohlmeier, P. et al. (2018). *Vitellogenin-like A*–associated shifts in social cue responsiveness regulate behavioral task specialization in an ant. *PLoS Biology, 16*, e2005747.

Masuko, K. (1966). Temporal division of labor among workers in the ponerine ant, *Amblyopone silvestrii. Sociobiology, 28*, 131–51.

Robinson, G. E. (1992). Regulation of division of labor in insect societies. *Annual Review of Entomology, 37*, 637–65.

SOME QUEEN ANTS FEED OFF LARVAL SECRETIONS

Børgesen, L. W., & Jensen, P. V. (1995). Influence of larvae and workers on egg production of queens of the pharaoh's ant, *Monomorium pharaonis* (L.). *Insectes Sociaux, 42*.

Cassill, D. L., & Vinson, S. B. (2007). Effects of larval secretions on queen fecundity in the fire ant. *Annals of the Entomological Society of America, 100*, 327–33.

Eder, J., & Rembold, H. (Eds.) (1986). *Chemistry and Biology of Social Insects.* Peperny.

Warner, M. R. et al. (2016). Late-instar ant worker larvae play a prominent role in colony-level caste regulation. *Insectes Sociaux, 63*, 575–83.

THE QUEENS OF *STIGMATOMMA SILVESTRII* FEED OFF THE HEMOLYMPH OF THEIR LARVAE

Masuko, K. (1986). Larval hemolymph feeding: A nondestructive parental cannibalism in the primitive ant *Amblyopone silvestrii. Behavioral Ecology and Sociobiology, 19*, 249–55.

Masuko, K. (2019). Larval hemolymph feeding and hemolymph taps in the ant *Proceratium itoi* (Hymenoptera: Formicidae). *Myrmecological News, 29*, 21–34.

Saux, C. et al. (2004). Dracula ant phylogeny as inferred by nuclear 28S rDNA sequences and implications for ant systematics (Hymenoptera: Formicidae: Amblyoponinae). *Molecular Phylogenetics and Evolution*, *33*, 457–68.

TROPHIC EGGS LAID BY WORKERS ARE EATEN BY THE QUEEN OR FED TO THE LARVAE

Dijkstra, M. B. et al. (2005). Self-restraint and sterility in workers of Acromyrmex and *Atta* leafcutter ants. *Insectes Sociaux*, *52*, 67–76.

Khila, A., & Abouheif, E. (2008). Reproductive constraint is a developmental mechanism that maintains social harmony in advanced ant societies. *Proceedings of the National Academy of Sciences USA*, *105*, 17884–89.

IN MANY SPECIES, WORKER ANTS LIVE LESS THAN A YEAR

Chapuisat, M., & Keller, L. (2002). Division of labour influences the rate of ageing in weaver ant workers. *Proceedings of the Royal Society B*, *269*, 909–13.

Giraldo, Y. M. et al. (2016). Lifespan behavioural and neural resilience in a social insect. *Proceedings of the Royal Society*, *283*.

Kohlmeier, P. et al. (2017). Intrinsic worker mortality depends on behavioural caste and the queens' presence in a social insect. *Science of Nature*, *104*, 34.

Kramer, B. H., & Schaible, R. (2013). Colony size explains the lifespan differences between queens and workers in eusocial Hymenoptera. *Biological Journal of the Linnean Society*, *109*, 710–72.

MOST ANT SPECIES HAVE A POISON STING, ONLY WOOD ANTS DO NOT

Blanchard, B. D., & Moreau, C. S. (2017). Defensive traits exhibit an evolutionary trade-off and drive diversification in ants. *Evolution*, *71*, 315–28.

Kazuma, K. et al. (2017). Combined venom gland transcriptomic and venom peptidomic analysis of the predatory ant *Odontomachus monticola*. *Toxins*, *9*, 23.

THE SCHMIDT STING PAIN INDEX INDICATES HOW PAINFUL AN INSECT'S STING IS

Evans, D. L., & Schmidt, J. O. (1990). *Insect Defenses: Adaptive Mechanisms and Strategies of Prey and Predators*. State University of New York Press.

Schmidt, J. O. et al. (1983). Hemolytic activities of stinging insect venoms. *Archives of Insect Biochemistry and Physiology*, *1*.

SOME SPECIES OF ANTS HAVE SMALLER MINOR WORKERS AND LARGER MAJOR WORKERS

Holley, J. A. C. et al. (2016). Subcaste-specific evolution of head size in the ant genus *Pheidole*. *Biological Journal of the Linnean Society*, *118*, 472–85.

Simola, D. F. et al. (2016). Epigenetic (re)programming of caste-specific behavior in the ant *Camponotus floridanus*. *Science*, *351*, aac6633.

Wheeler, D. E. (1991). The developmental basis of worker caste polymorphism in ants. *American Naturalist, 138* (5), 1218–38.

Wheeler, D. E., & Nijhout, H. F. (1981). Soldier determination in ants: New role for juvenile hormone. *Science, 213,* 361–63.

CHAPTER 2: THE BIRTH OF A COLONY

FEEDING DURING THE LARVAL STAGE DETERMINES WHICH LARVAE GO ON TO BECOME QUEENS

Bono, J. M., & Herbers, J. M. (2003). Proximate and ultimate control of sex ratios in *Myrmica brevispinosa* colonies. *Proceedings of the Royal Society B, 270,* 811–17.

Buschinger, A. (1990). Regulation of worker and queen formation in ants with special reference to reproduction and colony development. In Engels, W. (ed.). *Social Insects: An Evolutionary Approach to Castes and Reproduction.* Springer, 37–57.

Schwander, T. et al. (2010). Nature versus nurture in social insect caste differentiation. *Trends in Ecology & Evolution, 25,* 275–82.

Wheeler, D. E. (1986). Developmental and physiological determinants of caste in social Hymenoptera—evolutionary implications. *American Naturalist, 128,* 13–34.

MALES AND YOUNG QUEENS COUPLE UP ON THEIR NUPTIAL FLIGHTS

Hakala, S. M. et al. (2019). Evolution of dispersal in ants (Hymenoptera: Formicidae): A review on the dispersal strategies of sessile superorganisms. *Myrmecological News, 29,* 35–55.

Hart, A. G. et al. (2018). The spatial distribution and environmental triggers of ant mating flights: Using citizen-science data to reveal national patterns. *Ecography, 41,* 877–88.

Helms, J. A. (2016). The flight ecology of ants (Hymenoptera: Formicidae). *Myrmecological News, 26,* 19–30.

QUEEN ANTS ONLY MATE DURING THE NUPTIAL FLIGHT, OFTEN ONLY ONCE

Baer, B. (2016). Proximate and ultimate consequences of polyandry in ants (Hymenoptera: Formicidae). *Myrmecological News, 22,* 1–9.

Boomsma, J. J. (2009). Lifetime monogamy and the evolution of eusociality. *Philosophical Transactions of the Royal Society of London Series B, 364,* 3191–3207.

Boomsma, J. J., & Ratnieks, F. L. W. (1996). Paternity in eusocial Hymenoptera. *Philosophical Transactions of the Royal Society B, 351,* 947–75.

MALES NEVER WORK: THEY ARE ONLY REQUIRED FOR SEX, I.E., THE TRANSFER OF GENES

Boomsma, J. J. et al. (2005). The evolution of male traits in social insects. *Annual Review of Entomology, 50,* 395–420.

Heinze, J. (2016). The male has done his work—the male may go. *Current Opinion in Insect Science, 16*, 22–27.

THE FIGHTER MALES OF *CARDIOCONDYLA* REMAIN IN THE NEST AND MATE MULTIPLE TIMES

Anderson, C. et al. (2003). Live and let die: Why fighter males of the ant *Cardiocondyla* kill each other, but tolerate their winged rivals. *Behavioral Ecology, 14*, 54–62.

Frohschammer, S., & Heinze, J. (2009). Male fighting and "territoriality" within colonies of the ant *Cardiocondyla venustula. Naturwissenschaften, 96*, 159–63.

Heinze, J., & Hölldobler, B. (1993). Fighting for a harem of queens: Physiology of reproduction in *Cardiocondyla* male ants. *Proceedings of the National Academy of Sciences USA, 90*, 8412–14.

AFTER MATING, QUEENS OF MONOGYNOUS ANT SPECIES ESTABLISH A COLONY INDEPENDENTLY

Cahan, S. et al. (1998). An abrupt transition in colony founding behaviour in the ant *Messor pergandei. Animal Behavior, 55*, 1583–94.

Jeanson, R. & Fewell, J. H. (2008). Influence of the social context on division of labor in ant foundress associations. *Behavioral Ecology, 19*, 567–74.

Johnson, R. A. (2006). Capital and income breeding and the evolution of colony founding strategies in ants. *Insectes Sociaux, 53*, 316–22.

Tsuji, K., & Tsuji, N. (1996). Evolution of life history strategies in ants: Variation in queen number and mode of colony founding. *Oikos, 76*, 83–92.

IN POLYGYNOUS COLONIES, THE MATED YOUNG QUEEN RETURNS TO HER HOME NEST

Boulay, R. et al. (2004). The ecological benefits of larger colony size may promote polygyny in ants. *Journal of Evolutionary Biology, 27*, 2856–63.

Gill, R. J., & Arce, A. (2009). Polymorphic social organization in an ant. *Proceedings of the Royal Society B, 276*, 4423–31.

SOME YOUNG QUEENS FORCE THEIR WAY INTO THE COLONIES OF OTHER SPECIES AND TAKE THEM OVER

Buschinger, A. (2009). Social parasitism among ants: A review (Hymenoptera: Formicidae). *Myrmecological News, 12*, 219–35.

Chernenko, A. et al. (2013). Colony take-over and brood survival in temporary social parasites of the ant genus *Formica. Behavioral Ecology and Sociobiology, 67*, 727–35.

D'Ettorre, P. et al. (2002). Blending in with the crowd: Social parasites integrate into their host colonies using a flexible chemical signature. *Proceedings of the Royal Society B, 269*, 1911–18.

CHAPTER 3: EFFECTIVE ANARCHY

WORKERS DECIDE WHEN AND WHERE THE COLONY MOVES

Burns, D. D. R. et al. (2016). The effect of social information on the collective choices of ant colonies. *Behavioral Ecology, 27,* 1033–40.

Franks, N. R. et al. (2009). Speed versus accuracy in decision-making ants: Expediting politics and policy implementation. *Proceedings of the Royal Society B, 364,* 845–52.

McGlynn, T. P. (2012). The ecology of nest movement in social insects. *Annual Review of Entomology, 57,* 291–308.

Pratt, S. C. et al. (1992). Quorum sensing, recruitment, and collective decision-making during colony emigration by the ant *Leptothorax albipennis. Behavioral Ecology and Sociobiology, 52,* 117–27.

THE QUEEN'S BRAIN ATROPHIES AFTER THE NEST HAS BEEN ESTABLISHED

Gronenberg, W., & Liebig, J. (1999). Smaller brains and optic lobes in reproductive workers of the ant *Harpegnathos. Naturwissenschaften, 86,* 343–45.

Julian, G. E., & Gronenberg, W. (2002). Reduction of brain volume correlates with behavioral changes in queen ants. *Brain Behavior and Evolution, 60,* 152–64.

THE ACTIVITIY AND EXPERIENCES OF GENES DETERMINE AN ANT'S CHARACTER AND BEHAVIOR

Gospocic, J. et al. (2017). The neuropeptide corazonin controls social behavior and caste identity in ants. *Cell, 170,* 748.

Jandt, J. M. et al. (2014). Behavioural syndromes and social insects: Personality at multiple levels. *Biological Reviews, 89,* 48–67.

Jongepier, E., & Foitzik, S. (2016). Fitness costs of worker specialisation for ant societies. *Proceedings of the Royal Society B, 283,* 1,822.

Kohlmeier, P. et al. (2019). Gene expression is more strongly associated with behavioural specialisation than with age or fertility in ant workers. *Molecular Ecology, 28,* 658–70.

Ravary, F. et al. (2007). Individual experience alone can generate lasting division of labor in ants. *Current Biology, 17,* 1308–12.

WORKERS ARE MORE CLOSELY RELATED TO THEIR SISTERS THAN TO THE QUEEN

Sundström, L. (1994). Sex ratio bias, relatedness asymmetry and queen mating frequency in ants. *Nature, 367,* 266–67.

EUSOCIALITY REWARDS SELFLESS BEHAVIOR

Abbot, P. et al. (2011). Inclusive fitness theory and eusociality. *Nature, 471,* E1–E4.

Bourke, A. F. G. (2014). Hamilton's rule and the causes of social evolution. *Philosophical Transactions of the Royal Society B, 369.*

Pernu, T. K., & Helantera, H. (2019). Genetic relatedness and its causal role in the evolution of insect societies. *Journal of Bioscience*, *44*, 107.

Wilson, E. O., & Hölldobler, B. (2005). Eusociality: Origin and consequences. *Proceedings of the National Academy of Sciences USA*, *102*, 13367–71.

EUSOCIALITY DEVELOPED THROUGH COOPERATION BETWEEN SISTERS OR MOTHERS AND DAUGHTERS

Johnston, R. A., Cant, M. A., & Field, J. (2012). Sex-biased dispersal, haplodiploidy and the evolution of helping in social insects. *Proceedings of the Royal Society B*, *279*, 787–93.

Linksvayer, T. A., & Wade, M. J. (2005). The evolutionary origin and elaboration of sociality in the aculeate Hymenoptera: Maternal effects, sib-social effects, and heterochrony. *Quarterly Review of Biology*, *80*, 317–36.

CHAPTER 4: COMMUNICATIVE SENSUALITY

THE QUEENS OF *NEOPONERA INVERSA* CAN RECOGNIZE ONE ANOTHER CHEMICALLY

D'Ettorre, P., & Heinze, J. (2005). Individual recognition in ant queens. *Current Biology*, *15*, 2170–74.

ANTS RELY MORE ON THEIR SENSE OF SMELL THAN ON THEIR SENSE OF SIGHT

Gronenberg, W. (1999). Modality-specific segregation of input to ant mushroom bodies. *Brain Behavior and Evolution*, *54*, 85–95.

Knaden, M., & Graham, P. (2014). The sensory ecology of ant navigation: From natural environments to neural mechanisms. *Annual Review of Entomology*, *61*, 63–76.

ANTS SMELL AND TASTE USING THEIR FEELERS

Boroczky, K. et al. (2013). Insects groom their antennae to enhance olfactory acuity. *Proceedings of the National Academy of Sciences USA*, *110*, 3,615–3,620.

Draft, R. W. et al. (2018). Carpenter ants use diverse antennae sampling strategies to track odor trails. *Journal of Experimental Biology*, *221*, UNSP jeb185124.

Guerrieri, F. J., & d'Ettorre, P. (2010). Associative learning in ants: Conditioning of the maxilla-labium extension response in *Camponotus aethiops*. *Journal of Insect Physiology*, *56*, 88–92.

GLANDS SUCH AS THE DUFOUR'S GLAND PRODUCE SIGNALING SUBSTANCES, ALLOWING FOR COMPLEX CHEMICAL COMMUNICATION

Jackson, B. D., & Morgan, E. D. (1993). Insect chemical communication: Pheromones and exocrine glands in ants. *Chemoecology*, *4*: 125–44.

Jongepier, E. et al. (2015). The ecological success of a social parasite increases with manipulation of collective host behaviour. *Journal of Evolutionary Biology, 28,* 2152–62.

Mitra A. (2013). Function of the Dufour's gland in solitary and social Hymenoptera. Journal of Hymenopteran Research, *35,* 33–58.

EVERY COLONY OF ANTS HAS A NEST-SPECIFIC SCENT THANKS TO THEIR CUTICULAR HYDROCARBONS

Krasnec, M. O., & Breed, M. D. (2013). Colony-specific cuticular hydrocarbon profile in Formica argentea ants. *Journal of Chemical Ecology, 39,* 59–66.

Sturgis, S. J., & Gordon, D. M. (2012). Nestmate recognition in ants (Hymenoptera: Formicidae): a review. *Myrmecological News, 16,* 101–10.

ANTS HAVE SPECIFIC SCENTS ACCORDING TO CASTE AND JOB

Kleeberg, I. et al. (2017). The influence of slavemaking lifestyle, caste and sex on chemical profiles in Temnothorax ants: Insights into the evolution of cuticular hydrocarbons. *Proceedings of the Royal Society B, 284,* 1850.

Greene, M. J., & Gordon, D.M. (2003). Cuticular hydrocarbons inform task decisions. *Nature, 423,* 32.

WORKERS LAY SCENT TRAILS TO SOURCES OF FOOD

Czaczkes, T. J. et al. (2015). Trail pheromones: An integrative view of their role in social insect colony organization. *Annual Review of Entomology, 60,* 581–99.

Wilson, E. O. (1962). Chemical communication among workers of the fire ant Solenopsis saevissima. *Animal Behaviour, 10,* 134–64.

ANTS PERFORM TANDEM RUNS TO LEAD NEST-MATES TO NEW POTENTIAL NESTS AND SOURCES OF FOOD

Alleman, A. et al. (2019). Tandem-running and scouting behavior is characterized by up-regulation of learning and memory formation genes within the ant brain. *Molecular Ecology, 28,* 2342–59.

Franklin, E. L. (2014). The journey of tandem running: The twists, turns and what we have learned. *Insectes Sociaux, 61,* 1–8.

Franks, N. R., & Richardson, T. (2006). Teaching in tandem-running ants. *Nature, 439,* 153.

WORKERS EXCHANGE INFORMATION BY TOUCHING EACH OTHER'S FEELERS

Gill, K. P. et al. (2013). Density of antennal sensilla influences efficacy of communication in a social insect. *American Naturalist, 182,* 834–40.

Gordon, D. M., & Mehdiabadi, N. J. (1999). Encounter rate and task allocation in harvester ants. *Behavioral Ecology and Sociobiology, 45,* 370–77.

BURIED LEAFCUTTER ANTS USE VIBRATIONS TO CALL FOR HELP

Markl, H. (1965). Stridulation in Leaf-Cutting Ants. *Science, 149,* 1392–93.

Pielstrom, S., & Roces, F. (2012). Vibrational communication in the spatial organization of collective digging in the leaf-cutting ant Atta vollenweideri. *Animal Behaviour, 84,* 743–52.

CHAPTER 5: FINELY TUNED NAVIGATION

ANTS ORIENT THEMSELVES INSIDE THE NEST BY SMELL, TEMPERATURE, AND HUMIDITY

Heyman, Y. et al. (2019). Ants use multiple spatial memories and chemical pointers to navigate their nest. *iScience, 14,* 264–76.

Roemer, D. et al. (2018). Carbon dioxide sensing in the social context: Leaf-cutting ants prefer elevated CO2 levels to tend their brood. *Journal of Insect Physiology, 108,* 40–47.

ANTS BUILD ROADS WITH SCENT-BASED GUIDANCE SYSTEMS

Gordon, D. M, (2019). Local regulation of trail networks of the arboreal Turtle Ant, *Cephalotes goniodontus. American Naturalist, 190,* E156–E169.

Steck, K. (2012). Just follow your nose: Homing by olfactory cues in ants. *Current Opinion in Neurobiology, 22,* 231–35.

Wilson, E. O. (1959). Source and possible nature of the odor trail of fire ants. *Science, 129,* 643–44.

THE AFRICAN STINK ANT *PACHYCONDYLA TARSATA* ORIENTS ITSELF VISUALLY TO ITS SURROUNDINGS

Baader, A. P. (1996). The significance of visual landmarks for navigation of the tropical ant, *Paraponera clavata. Insectes Sociaux, 43,* 435–50.

Hölldobler, B. (1980). Canopy orientation: A new kind of orientation in ants. *Science, 210,* 86–88.

Oliveira, P. S., & Hölldobler, B. (1989). Orientation and communication in the Neotropical ant *Odontomachus bauri* Emery. *Ethology, 83,* 154–66.

THE POSITION OF THE SUN REVEALS WHICH WAY TO GO

Schwarz, S. et al. (2017). How ants use vision when homing backward. *Current Biology, 27,* 401–07.

THE DIRECTION OF THE LIGHT'S POLARIZATION REVEALS THE SUN'S POSITION EVEN ON CLOUDY DAYS

Lehrer, M. (Ed.) (1997). *Orientation and Communication in Arthropods.* Birkhäuser.

Wehner, R. (2003). Desert ant navigation: How miniature brains solve complex tasks. *Journal of Comparative Physiology A, 189,* 579–88.

Wehner, R., & Müller, M. (2006). The significance of direct sunlight and polarized skylight in the ant's celestial system of navigation. *Proceedings of the National Academy of Sciences USA*, *103*, 12575–79.

Zeil, J. et al. (2014). *Polarisation vision in ants, bees and wasps*. Springer.

THE DESERT ANT *CATAGLYPHIS* HAS AN INTERNAL NAVIGATION SYSTEM WITH A HOMING FUNCTION

Knaden, M., & Wehner, R. (2006). Ant navigation: Resetting the path integrator. *Journal of Experimental Biology*, *209*, 26–31.

Wittlinger, M. et al. (2006). The ant odometer: Stepping on stilts and stumps. *Science*, *312*, 1965–67.

ANTS REMEMBER THE PATH TO TAKE BY MEANS OF OPTIC FLOW

Pfeffer, S. E., & Wittlinger, M. (2016). Optic flow odometry operates independently of stride integration in carried ants. *Science*, *353*, 1155–57.

Ronacher, B., & Wehner, R. (1995). Desert ants *Cataglyphis fortis* use self-induced optic flow to measure distances travelled. *Journal of Comparative Physiology A*, *177*, 21–27.

Seidl, T. et al. (2006). Desert ants: is active locomotion a prerequisite for path integration? *Journal of Comparative Physiology A*, *192*, 1125–31.

Zollikofer, C. et al. (1995). Optical scaling in conspecific *Cataglyphis* ant. *Journal of Experimental Biology*, *198*, 1637–46.

SOME ANTS HAVE A MAGNETIC COMPASS

Banks, A. N., & Srygley, R. B. (2003). Orientation by magnetic field in leaf-cutter ants, *Atta colombica* (Hymenoptera: Formicidae). *Ethology*, *109*, 835–46.

Fleischmann, P. N. (2018). The geomagnetic field is a compass cue in *Cataglyphis* ant navigation. *Current Biology*, *28*, 1440–44.

CHAPTER 6: SAVAGE HORDES

ARMY ANTS HUNT IN LARGE SQUADRONS

Chadab, R., & Rettenmeyer, C. W. (1975). Mass recruitment by army ants. *Science*, *188*, 1124–25.

Couzin, I. D., & Franks, N. R. (2003). Self-organized lane formation and optimized traffic flow in army ants. *Proceedings of the Royal Society B*, *270*, 139–46.

Peters, M. K. et al. (2013). Spatial variation in army ant swarm raiding and its potential effect on biodiversity. *Biotropica*, *45*, 54–62.

ARMY ANT BIVOUACS ARE BUILT WITH THE LIVE BODIES OF WORKERS

Baudier, K. M. et al. (2019). Plastic collective endothermy in a complex animal society (army ant bivouacs: Eciton burchellii parvispinum). *Ecography*, *42*, 730–39.

Berghoff, S. M. et al. (2002). Nesting habits and colony composition of the hypogaeic army ant *Dorylus (Dichthadia) laevigatus* Fr. Smith. *Insectes Sociaux*, *49*, 380–87.

Franks, N. R. (1989). Thermoregulation in army ant bivouacs. *Physiological Entomology*, *14*, 397–404.

Jackson, W. B. (1957). Microclimatic patterns in the army ant bivouac. *Ecology*, 38, 276–85.

Schneirla, T. C. et al. (1954). The bivouac or temporary nest as an adaptive factor in certain terrestrial species of army ants. *Ecological Monographs*, *24*, 269–96.

NUMEROUS GUESTS LIVE INSIDE ARMY ANT NESTS

Rettenmeyer, C. W. et al. (2011.) The largest animal association centered on one species: the army ant *Eciton burchellii* and its more than 300 associates. *Insectes Sociaux*, *58*, 281–92.

von Beeren, C. et al. (2018). Chemical and behavioral integration of army ant-associated rove beetles—a comparison between specialists and generalists. *Frontiers in Zoology*, *15*, 8.

Witte, V. et al. (2008). Symbiont microcosm in an ant society and the diversity of interspecific interactions. *Animal Behaviour*, *76*, 1477–86.

IN THE EVENT OF FLOODING, ANTS FORM RAFTS TO FLOAT ON TOP OF THE WATER

Mlot, N. J. et al. (2011). Fire ants self-assemble into waterproof rafts to survive floods. *Proceedings of the National Academy of Sciences USA*, *108*, 7669–73.

Purcell, J. et al. (2014). Ant brood function as life preservers during floods. *PLoS One*, *9*, e89211.

Reid, C. R. et al. (2015). Army ants dynamically adjust living bridges in response to a cost-benefit trade-off. *Proceedings of the National Academy of Sciences USA*, *112*, 15113–18.

ARMY ANTS ALTERNATE BETWEEN A VAGABOND EXISTENCE AND A MORE SETTLED ONE

O'Donnell, S. et al. (2009). Species and site differences in Neotropical army ant emigration behaviour. *Ecological Entomology*, *34*, 476–82.

Garnier, S., Kronauer, D. J. C. (2017). The adaptive significance of phasic colony cycles in army ants. *Journal of Theoretical Biology*, *428*, 43–47.

Willson, S. K. et al. (2001). Spatial movement optimization in Amazonian *Eciton burchellii* army ants. *Insectes Sociaux*, *58*, 325–34.

ARMY ANTS MATE INSIDE THE NEST

Kronauer, D. J. C. (2009). Recent advances in army ant biology (Hymenoptera: Formicidae). *Myrmecological News*, *12*, 51–56.

Kronauer, D. J. C. et al. (2006). A reassessment of the mating system characteristics of the army ant *Eciton burchellii*. *Naturwissenschaften*, *93*, 402–06.

Kronauer, D. J. C. et al. (2007). The evolution of multiple mating in army ants. *Evolution, 61,* 413–22.

IF THE COLONY GROWS TOO LARGE, IT SPLITS

Franks, N. R., & Hölldobler, B. (1987). Sexual competition during colony reproduction in army ants. *Biological Journal of the Linnean Society, 30,* 229–43.

Kronauer, D. J. C. et al. (2004). Extreme queen-mating frequency and colony fission in African army ants. *Molecular Ecology, 13,* 2381–88.

PRIMITIVE PEOPLES USE BITING ANTS TO SUTURE WOUNDS

Boswell, G. P. et al. (2001). Arms races and the evolution of big fierce societies. *Proceedings of the Royal Society B, 268,* 1723–30.

Davies, H. E. (2019). Leaf-cutter ants in wound closure. *Wilderness & Environmental Medicine, 30*(4).

Schiappa, J., & Van Hee, R. (2016). From ants to staples: History and ideas concerning suturing techniques. *Acta Chirurgica Belgica, 112,* 395–402.

THE MOFU IN CAMEROON FIGHT TERMITES WITH ARMY ANTS

Seignobos, C. (2007). Jaglavak, prince of insects. www.pbs.org/wgbh/nova/article/jaglavak-prince-of-insects.

Seignobos, C. et al. (1996). Les Mofus et leur insectes. *Journal d'agriculture traditionnelle et de botanique appliquée, 38,* 125–87.

CHAPTER 7: A GARDEN FOR A CITY OF MILLIONS

YOUNG QUEENS TAKE FUNGAL TISSUE FROM THEIR MOTHER NEST AS A DOWRY

Green, A. M. (2002). Extensive exchange of fungal cultivars between sympatric species of fungus-growing ants. *Molecular Ecology, 11,* 191–95.

Marti, H. E. et al. (2015). Foundress queen mortality and early colony growth of the leafcutter ant, *Atta texana* (Formicidae, Hymenoptera). *Insectes Sociaux, 62,* 357–63.

Meirelles, L. A. et al. (2016). Bacterial microbiomes from vertically transmitted fungal inocula of the leaf-cutting ant *Atta texana*. *Environmental Microbiology Reports, 8,* 630–40.

COLONIES OF LEAFCUTTER ANTS BOAST WORKERS OF DIFFERENT SIZES WITH DIFFERENT TASKS

Burd, M. (2000). Body size effects on locomotion and load carriage in the highly polymorphic leaf-cutting ants *Atta colombica* and *Atta cephalotes*. *Behavioral Ecology, 11,* 125–31.

Constant, N. et al. (2012). The effects of genotype, caste, and age on foraging performance in leaf-cutting ants. *Behavioral Ecology, 23,* 1284–88.

Hughes, W. O. H. et al. (2003). Worker caste polymorphism has a genetic basis in *Acromyrmex* leaf-cutting ants. *Proceedings of the National Academy of Sciences USA, 100,* 9394–97.

Wills, B. D. et al. (2018). Correlates and consequences of worker polymorphism in ants. *Annual Review of Entomology, 63,* 575–98.

CEMENT-FILLED LEAFCUTTER NESTS REVEAL THEIR SIZE AND STRUCTURE

Jonkman, J. C. M. (1980). The external and internal structure and growth of nests of the leaf-cutting ant *Atta vollenweideri* Forel, 1893 (Hym.: Formicidae). *Journal of Applied Entomology, 89,* 158–73.

Moreira, A. A. et al. (2004). External and internal structure of *Atta bisphaerica* Forel (Hymenoptera: Formicidae) nests. *Journal of Applied Entomology, 128,* 204–11.

LEAFCUTTER ANTS CULTIVATE FUNGUS

Bass, M., & Cherrett, J. M. (1995). Fungal hyphae as a source of nutrients for the leaf-cutting ant *Atta sexdens. Physiological Entomology, 20,* 1–6.

Hölldobler, B., & Wilson, E. O. (2010). *The Leafcutter Ants: Civilization by Instinct.* W. W. Norton & Company.

North, R. D. et al. (1997). Evolutionary aspects of ant-fungus interactions in leaf-cutting ants. *Trends in Ecology & Evolution, 12,* 386–89.

Poulsen, M., & Boomsma J. J. (2005). Mutualistic fungi control crop diversity in fungus-growing ants. *Science, 307,* 741–44.

ANTS USE ANTIFUNGALS TO PROTECT THE FUNGUS ON WHICH THEY DEPEND

Currie, C. R. et al. (2003). Ancient tripartite coevolution in the attine ant-microbe symbiosis. *Science, 299,* 386–88.

Haeder, S. et al. (2009). Candicidin-producing *Streptomyces* support leaf-cutting ants to protect their fungus garden against the pathogenic fungus *Escovopsis. Proceedings of the National Academy of Sciences USA, 106,* 4742–46.

Santos, A. V. et al. (2004). Occurrence of the antibiotic producing bacterium *Burkholderia* sp. in colonies of the leaf-cutting ant *Atta sexdens rubropilosa. FEMS Microbiology Letters, 239*(2), 319–23.

LEAFCUTTER ANTS CAUSE GREAT DAMAGE ON AGRICULTURAL PLANTATIONS

Blanton, C. M., & Ewel, J. J. (1985). Leaf-cutting ant herbivory in successional and agricultural tropical ecosystems. *Ecology, 66,* 861–69.

Costa, A. N. et al. (2018). Do an ecosystem engineer and environmental gradient act independently or in concert to shape juvenile plant communities? Tests with the leaf-cutter ant *Atta laevigata* in a Neotropical savanna. *PeerJ,* e5612.

Montoya-Lerma, J. et al. (2012). Leaf-cutting ants revisited: Towards rational management and control. *International Journal of Pest Management*, *58*, 225–47.

ANTS SUCCESSFULLY DEFEND THEIR PLANTS AGAINST LEAFCUTTER ANTS

Adams, E. S. (1994). Territory defense by the ant *Azteca trigona*: Maintenance of an arboreal ant mosaic. *Oecologia*, *97*, 202–08.

Schmidt, M., & Dejean, A. (2018). A Dolichoderine ant that constructs traps to ambush prey collectively: Convergent evolution with a Myrmicine Genus. *Biological Journal of the Linnean Society*, *124*, 41–46.

Wetterer, J. (1994). Attack by *Paraponera clavata* prevents herbivory by the leaf-cutting ant, *Atta cephalotes*. *Biotropica*, *26*, 462–65.

SCUTTLE FLIES CATCH A RIDE ON TRANSPORT WORKERS

Elizalde, L. et al. (2018). Spatial and temporal variation in host-parasitoid interactions: Leafcutter ant hosts and their phorid parasitoids. *Ecological Entomology*, *43*, 114–125.

Feener, D. H., & Brown, B. V. (1993). Oviposition behavior of an ant-parasitizing fly, *Neodohrniphora curvinervis* (Diptera: Phoridae), and defense behavior by its leaf-cutting ant host *Atta cephalotes* (Hymenoptera: Formicidae). *Journal of Insect Behavior*, *6*, 675–688.

Orr, M. R. (1992). Parasitic flies (Diptera: Phoridae) influence foraging rhythms and caste division of labor in the leaf-cutter ant, *Atta cephalotes* (Hymenoptera: Formicidae). *Behavioral Ecology and Sociobiology*, *30*, 395–402.

CHAPTER 8: A TREEHOUSE FOR THE NATION

CAMPONOTUS SCHMITZI LIVES INSIDE PITCHER PLANTS, STEALING THEIR PREY

Bonhomme, V. et al. (2011). The plant-ant *Camponotus schmitzi* helps its carnivorous host-plant *Nepenthes bicalcarata* to catch its prey. *Journal of Tropical Ecology*, *27*, 15–24.

Thornham, D. G. et al. (2011). Setting the trap: cleaning behaviour of *Camponotus schmitzi* ants increases long-term capture efficiency of their pitcher plant host, *Nepenthes bicalcarata*. *Functional Ecology*, *26*, 11–19.

AZTECA ANTS KEEP *CATALPA* TREES FREE FROM EPIPHYTES

Janzen, D. H. (1969). Allelopathy by myrmecophytes: The ant azteca as an allelopathic agent of Cecropia. *Ecology*, *50*, 147–53.

Marting, P. R. et al. (2018). Colony personality and plant health in the Azteca-Cecropia mutualism. *Behavioral Ecology*, *29*, 264–71.

Mayer, V. E. et al. (2014). Current issues in the evolutionary ecology of ant-plant symbioses. *New Phytologist*, *202*, 749–64.

THE BULL HORN ACACIA MAKES *PSEUDOMYRMEX FERRUGINEA* DEPENDENT ON ITS NECTAR

Heil, M. et al. (2004). Main nutrient compounds in food bodies of Mexican Acacia ant-plants. *Chemoecology, 14,* 45–52.

Heil, M. et al. (2005). Postsecretory hydrolysis of nectar sucrose and specialization in ant/plant mutualism. *Science, 308,* 560–63.

Heil, M. et al. (2014). Partner manipulation stabilises a horizontally transmitted mutualism. *Ecology Letters, 17,* 185–92.

Pennisi, E. (2005). Sucrose-free sips suit acacia ants. *Science, 308,* 481–82.

THE LEMON ANT *MYRMELACHISTA SCHUMANNI* CREATES MONOCULTURES IN THE JUNGLE, KNOWN AS DEVIL'S GARDENS

Baez, S. et al. (2016). Ant mutualism increases long-term growth and survival of a common Amazonian tree. *American Naturalist, 188,* 567–75.

Frederickson, M. E. et al. (2005). "Devil's gardens" bedevilled by ants. *Nature, 437,* 495–96.

Frederickson, M. E., & Gordon, D. M. (2007). The devil to pay: a cost of mutualism with *Myrmelachista schumanni* ants in "devil's gardens" is increased herbivory on *Duroia hirsuta* trees. *Proceedings of the Royal Society B, 247,* 1117–23.

Salas-Lopez, A. et al. (2016). The discovery of devil's gardens: an ant-plant mutualism in the cloud forests of the Eastern Amazon. *Journal of Tropical Ecology, 32,* 264–68.

CAMPONOTUS FEMORATUS BUILDS HANGING GARDENS IN THE TREETOPS

Cereghino, R. et al. (2011). Ant-plant mutualisms promote functional diversity in phytotelm communities. *Functional Ecology, 25,* 954–63.

Orivel, J., & Dejean, A. (1999). Selection of epiphyte seeds by ant-garden ants. *Écoscience, 6,* 51–55.

Seidel, J. L. et al. (1990). Neotropical ant gardens. *Journal of Chemical Ecology, 16,* 1791–1816.

GREEN WEAVER ANTS WEAVE LEAVES TOGETHER WITH SILK TO BUILD NESTS

Bochynek, T., & Robson, S. K. A. (2014). Physical and biological determinants of collective behavioural dynamics in complex systems: Pulling chain formation in the nest-weaving ant *Oecophylla smaragdina. PLoS One, 9,* e95112.

Cole, A. C. et al. (1948.) A study of the weaver ant, *Oecophylla smaragdina* (Fab.). *American Midland Naturalist, 39,* 641–651.

Hölldobler, B. (1983). Territorial behavior in the green tree ant (*Oecophylla smaragdina*). *Biotropica, 15,* 241–250.

CHAPTER 9: MILKING IT: ANTS AND THEIR LIVESTOCK

ANTS KEEP APHIDS AND OTHER SAP-SUCKING INSECTS AS
LIVESTOCK

Cushman, J. H., & Addicott, J. F. (1989). Intra- and interspecific competition for mutualists: ants as a limited and limiting resource for aphids. *Oecologia, 79*, 315–21.

Detrain, C. et al. (2010). Aphid-ant mutualism: how honeydew sugars influence the behaviour of ant scouts. *Physiological Entomology, 35*, 168–74.

Fischer, M. K. et al. (2003). Competition for mutualists in an ant-homopteran interaction mediated by hierarchies of ant attendance. *Oikos, 92*, 531–41.

Völkl, W. et al. (1999). Ant-aphid mutualisms: The impact of honeydew production and honeydew sugar composition on ant preferences. *Oecologia, 118*, 483–91.

ANTS AND THEIR APHIDS HAVE ADAPTED TO EACH OTHER

Mani, M., & Shivaraju, C. (Eds.) (2016). *Mealybugs and their management in agricultural and horticultural crops.* Springer.

Maschwitz, U., & Hänel, H. (1985). The migrating herdsman *Dolichoderus* (*Diabolus*) *cuspidatus*: An ant with a novel mode of life. *Behavioral Ecology and Sociobiology, 17*, 171–84.

Ross, L., & Shuker, D. M. (2009). Scale insects. *Current Biology, 19*, R184–R186.

Shingleton, A. W. (2007). The origin of a mutualism: A morphological trait promoting the evolution of ant-aphid mutualisms. *Evolution, 59*, 921–26.

Way, M. J. (1963). Mutualism between ants and honeydew-producing Homoptera. *Annual Review of Entomology, 8*, 307–44.

THE BLACK GARDEN ANT EXPLOITS THE BLACK BEAN APHID

Banks, C. J. (1958). Effects of the Ant, *Lasius niger* (L.), on the Behaviour and Reproduction of the Black Bean Aphid, *Aphis fabae* Scop. *Bulletin of Entomological Research, 49*, 701–14.

Banks, C. J., & Nixon, H. L. (1958). Effects of the Ant, *Lasius Niger* L., on the Feeding and Excretion of the Bean Aphid, *Aphis Fabae* Scop. *Journal of Experimental Biology, 35*, 703–11.

Offenberg, J. (2001). Balancing between mutualism and exploitation: The symbiotic interaction between *Lasius* ants and aphids. *Behavioral Ecology and Sociobiology, 49*, 304–10.

THE LARVAL FORM OF THE APHID *PARACLETUS CIMICIFORMIS* EATS
THE LARVAE OF THE PAVEMENT ANT *TETRAMORIUM CAESPITUM*

Depa, Ł. et al. (2017). Do ants drive speciation in aphids? A possible case of ant-driven speciation in the aphid genus *Stomaphis* Walker (Aphidoidea, Lachninae). *Zoological Journal of the Linnean Society, 179*, 41–61.

Salazar, A. et al. (2015). Aggressive mimicry coexists with mutualism in an aphid. *Proceedings of the National Academy of Sciences USA, 112,* 1101–06.

GOSSAMER-WINGED BUTTERFLIES ARE PROTECTED AND CARED FOR BY ANTS AS CATERPILLARS

Fiedler, K. (2012). The host genera of ant-parasitic Lycaenidae butterflies: A review. *Psyche, 2012.*

Henning, S. F. (1983). Chemical communication between lycaenid larvae (Lepidoptera: Lycaenidae) and ants (Hymenoptera: Formicidae). *Journal of the Entomological Society of Southern Africa, 46,* 341–66.

Pierce, N. et al. (2002). The Ecology and Evolution of Ant Association in the Lycaenidae (Lepidoptera). *Annual Review of Entomology, 47,* 733–71.

ANT GUESTS PROTECT THEMSELVES OR ADJUST THEIR SCENT TO GO UNDETECTED

Agrain, F. A. et al. (2015). Leaf beetles are ant-nest beetles: the curious life of the juvenile stages of case-bearers (*Coleoptera, Chrysomelidae, Cryptocephalinae*). *Zookeys, 547,* 133–64.

Erber, D. (1988). Biology of *Camptosomata Clytrinae - Cryptocephalinae - Chlamisinae - Lamprosomatinae.* In Jolivet, P., Petitpierre, E., & Hsiao, T.H. (Eds.) *Biology of Chrysomelidae.* (Vol. 42). Springer.

Hlaváč, P. (2005). Revision of the myrmecophilous genus *Lomechusa* (Coleoptera: Staphylinidae: Aleocharinae). *Sociobiology, 46,* 203–50.

Hölldobler, B. et al. (2018). Behavior and exocrine glands in the myrmecophilous beetle *Lomechusoides strumosus* (Fabricius, 1775) (formerly called *Lomechusa strumosa*) (Coleoptera: Staphylinidae: Aleocharinae). *PLoS One, 13,* e0200309.

Witek, M. et al. (2014). *Myrmica* ants host highly diverse parasitic communities: From social parasites to microbes. *Insectes Sociaux, 61,* 307.

CHAPTER 10: ON PARASITES AND SLAVE-MAKERS

TEMNOTHORAX AMERICANUS ENSLAVES *T. LONGISPINOSUS*

Foitzik, S. et al. (2001). Coevolution in host-parasite systems: Behavioral strategies of slavemaking ants and their hosts. *Proceedings of the Royal Society B, 268,* 1139–46.

Kaur, R. et al. (2019). Ant behaviour and brain gene expression of defending hosts depend on the ecological success of the intruding social parasite. *Philosophical Transactions of the Royal Society B, 374,* 1769.

Pamminger, T. et al. (2012). Raiders from the sky: Slavemaker founding queens select for aggressive host colonies. *Biology Letters, 8,* 748–50.

Wesson, L. G. (1939). Contributions to the natural history of *Harpagoxenus americanus*. *Transactions of the American Entomological Society*, *65*, 97–122.

HONEYPOT ANTS AMBUSH AND ROB HARVESTER ANTS
Hölldobler, B. (1968). Food robbing in ants, a form of interference competition. *Oecologia*, *69*, 12–15.

THE THIEF ANT BREAKS INTO THE NESTS OF OTHER SPECIES
Hölldobler, B. (1973). Chemical strategy during foraging in *Solenopsis fugax* Latr and *Monomorium pharaonis* L. *Oecologia*, 11, 371–80.

MEGALOMYRMEX SYMMETOCHUS EXPLOIT SERICOMYRMEX ANTS, BUT PROTECT THEM AGAINST RAIDS BY GNAMPTOGENYS HARTMANI
Adams, R. M. M. et al. (2013). Chemically armed mercenary ants protect fungus-farming societies. *Proceedings of the National Academy of Sciences USA*, *110*, 15752–57.

T. MINUTISSIMUS NO LONGER HAS WORKERS AND HAS T. CURVISPINOSUS DO ALL THE HARD WORK
Johnson, C. A. et al. (2008). Stealth and reproductive dominance in a rare parasitic ant. *Animal Behavior*, *76*, 1965–76.

QUEENS OF T. SCHNEIDERI RIDE THEIR HOST ANTS
Bourke, A. F. G., & Franks N. R. (1991). Alternative adaptions, sympatric speciation and the evolution of parasitic, inquiline ants. *Biological Journal of the Linnean Society*, *43*, 157–78.
Kutter, H. (1950). Über eine neue, extrem parasitische Ameise. *Mitteilungen der Schweizerischen Entomologischen Gesellschaft*, *23*, 81–94.
Stumper, R. (1951). *Teleutomyrmex schneideri* Kutter (Hym. Formicid.). II. Mitteilung. Über die Lebensweise der neuen Schmarotzerameise. *Mitteilungen der Schweizerischen Entomologischen Gesellschaft*, *24*, 129–52.

MYRMECOCYSTUS HONEYPOT ANTS HAVE SHOW FIGHTS OR ENSLAVE THE COMPETITION
Hölldobler, B. (1976). Tournaments and slavery in a desert ant. *Science*, *192* (4242), 912–14.

SLAVE-MAKER QUEENS CREEP INTO OTHER NESTS BY ESCHEWING THEIR SCENT OR IN DISGUISE
D'Ettore, P. et al. (2002). Blending in with the crowd: Social parasites integrate into their host colonies using a flexible chemical signature. *Proceedings of the Royal Society B*, *269*, 1911–18.
D'Ettore, P. et al. (2000). Sneak in or repel your enemy: Dufour's gland repellent as a strategy for successful usurpation in the slave-maker *Polyergus rufescens*. *Chemoecology*, *10*, 135–42.

Johnson, C. A. et al. (2001). Changes in the cuticular hydrocarbon profile of the slave-maker ant queen, *Polyergus breviceps* emery, after killing a Formica host queen (Hymenoptera: Formicidae). *Journal of Chemical Ecology, 27,* 1787–1804.

Kleeberg, I. et al. (2017). The influence of slavemaking lifestyle, caste and sex on chemical profiles in *Temnothorax* ants: Insights into the evolution of cuticular hydrocarbons. *Proceedings of the Royal Society B, 284,* 1850.

POLYERGUS SLAVE-MAKERS RELY ON BRUTE FORCE DURING RAIDS

Topoff, H. et al. (1989). Behavioral adaptations for raiding in the slave-making ant, *Polyergus breviceps. Journal of Insect Behavior, 2,* 545–56.

MANY SLAVE-MAKERS USE SECRETIONS FROM THEIR DUFOUR'S GLAND TO INDUCE THEIR HOSTS TO ATTACK ONE ANOTHER

Allies, A. B. et al. (1986). Propaganda substances in the cuckoo ant *Leptothorax kutteri* and the slave-maker *Harpagoxenus sublaevis. Journal of Chemical Ecology, 12,* 1285–93.

Bauer, S. et al. (2009). Fight or flight? A geographic mosaic in host reaction and potency of a chemical weapon in the social parasite *Harpagoxenus sublaevis. Behavioral Ecology and Sociobiology, 64,* 45–56.

Brandt, M. et al. (2006). Dufour's gland secretion as a propaganda substance in the slavemaking ant *Protomognathus americanus. Insectes Sociaux, 53,* 291–99.

Jongepier, E. et al. (2015). The ecological success of a social parasite increases with manipulation of collective host behaviour. *Journal of Evolutionary Biology, 28,* 2152–62.

Savolainen, R., & Deslippe, R. J. (1996). Facultative and obligate slavery in formicine ants: Frequency of slavery, and proportion and size of slaves. *Biological Journal of the Linnean Society, 57,* 47–58.

REGULAR, DESTRUCTIVE RAIDS BY SLAVE-MAKERS HARM THE HOST POPULATIONS

Foitzik, S., & Herbers, J. M. (2001). Colony structure of a slavemaking ant: II. Frequency of slave raids and impact on the host population. *Evolution, 55,* 316–23.

Foitzik, S. et al. (2009). Locally-adapted social parasite affects density, social structure and life history of its ant hosts. *Ecology, 90,* 1195–1206.

THE "NINJA ANT" *TEMNOTHORAX PILAGENS* CAN ROB HOST NESTS ALMOST UNDETECTED

Kleeberg, I., & Foitzik S. (2016). The placid slavemaker: Avoiding detection and conflict as an alternative, peaceful raiding strategy. *Behavioral Ecology and Sociobiology, 70,* 27–39.

Seifert, B. et al. (2014). *Temnothorax pilagens* sp. n.—a new slave-making species of the tribe Formicoxenini from North America (Hymenoptera, Formicidae). *Zookeys, 368,* 65–77.

GENETIC VARIATIONS TURN ANTS INTO SLAVE-MAKERS,
ENSLAVING CLOSELY RELATED SPECIES

Alleman, A. et al. (2018). Comparative analyses of co-evolving host-parasite associations reveal unique gene expression patterns underlying slavemaker raiding and host defensive phenotypes. *Scientific Reports*, *8*, 1951.

Feldmeyer, B. et al. (2017). Species-specific genes under selection characterize the co-evolution of slavemaker and host lifestyles. *BMC Evolutionary Biology*, *17*, 237.

HOST SPECIES DEVELOP DEFENSIVE STRATEGIES AGAINST
ATTACKS BY SLAVE-MAKERS

Jongepier, E. et al. (2014). Collective defense portfolios of ant hosts shift with social parasite pressure. *Proceedings of the Royal Society B*, *281*, 1791.

Kleeberg, I. et al. (2014). Forewarned is forearmed: Aggression and information use determine fitness costs of slave raids. *Behavioral Ecology*, *25*, 1058–63.

Pamminger, T. et al. (2011). Increased host aggression as an induced defence against slavemaking ants. *Behavioral Ecology*, *22*, 255–60.

SOME ENSLAVED ANTS REBEL AND KILL THE SLAVE-MAKERS'
OFFSPRING

Achenbach, A., Foitzik, S. (2009). First evidence for slave rebellion: Enslaved ant workers systematically kill the brood of their social parasite *Protomognathus americanus*. *Evolution*, *63*, 1068–75.

Achenbach, A. et al. (2010). Brood exchange experiments and chemical analyses shed light on slave rebellion in ants. *Behavioral Ecology*, *21*, 948–56.

Metzler, D. et al. (2016). The influence of space and time on the evolution of altruistic defense: the case of ant slave rebellion. *Journal of Evolutionary Biology*, *29*, 874–86.

Pamminger, T. et al. (2014). Oh sister, where art thou? Spatial population structure and the evolution of an altruistic defence trait. *Journal of Evolutionary Biology*, *27*, 2443–56.

CHAPTER 11: PHYSICIAN, HEAL THYSELF

THE NEMATODE *MYRMECONEMA NEOTROPICUM* TURNS ANTS
INTO RED BERRIES

Dattilo, W. et al. (2013). The Geographic Distribution of Parasite-Induced Fruit Mimicry in *Cephalotes atratus* (Formicidae: Myrmicinae). *Journal of Parasitology*, *99*, 155–57.

Poinar, G. Jr. (2012). Nematode parasites and associates of ants: Past and present. *Psyche*, *2012*.

Robin, M. V. et al. (2012). Exoskeletal thinning in *Cephalotes atratus* ants (Hymenoptera: Formicidae) parasitized by *Myrmeconema neotropicum* (Nematoda: Tetradonematidae). *Journal of Parasitology, 98,* 226–28.

Yanoviak, S. P. et al. (2008). Parasite-induced fruit mimicry in a tropical canopy ant. *American Naturalist, 171,* 536–44.

THE TAPEWORM *ANOMOTAENIA BREVIS* EXTENDS THE LIFE OF THE AFFLICTED ANT AND MANIPULATES ITS BEHAVIOR

Beros, S. et al. (2015). The parasite's long arm: a tapeworm parasite induces behavioural changes in uninfected group members of its social host. *Proceedings of the Royal Society B, 282.*

Beros, S. et al. (2019). Parasitism and queen presence interactively shape worker behaviour and fertility in an ant host. *Animal Behavior, 148,* 63–70.

Feldmeyer, B. et al. (2016). Gene expression patterns underlying parasite-induced alterations in host behaviour and life history. *Molecular Ecology, 25,* 648–60.

Scharf, I. et al. (2012). Ant societies buffer individual-level effects of parasite infections. *American Naturalist, 180,* 671–83.

Trabalon, M. et al. (2000). Modification of morphological characters and cuticular compounds in worker ants *Leptothorax nylanderi* induced by endoparasites *Anomotaenia brevis. Journal of Insect Physiology, 46*(2), 169–78.

PARASITIC FUNGI CAN TURN ANTS INTO ZOMBIES

Araújo, J. P. M. et al. (2018). Zombie-ant fungi across continents: 15 new species and new combinations within *Ophiocordyceps.* I. Myrmecophilous hirsutelloid species. *Studies in Mycology, 90,* 119–60.

Fredericksen, M. A. et al. (2017). Three-dimensional visualization and a deep-learning model reveal complex fungal parasite networks in behaviorally manipulated ants. *Proceedings of the National Academy of Sciences USA, 114,* 12590–95.

Hughes, D. P. et al. (2011). Ancient death-grip leaf scars reveal ant-fungal parasitism. *Biology Letters, 7,* 67–70.

Kobmoo, N. et al. (2019). Population genomics revealed cryptic species within host-specific zombie-ant fungi (*Ophiocordyceps unilateralis*). *Molecular Phylogenetics and Evolution, 140.*

Lachaud, J-P. et al. (2013). Ants and their parasites 2013. *Psyche.*

Loreto, R. G., & Hughes, D. P. (2019). The metabolic alteration and apparent preservation of the zombie ant brain. *Journal of Insect Physiology, 118.*

THE LIVER FLUKE *DICROCOELIUM DENDRITICUM* ALSO FORCES ANTS TO DO ITS BIDDING

Botvenik, C. E. et al. (2016). Relative effects of temperature, light, and humidity on clinging behavior of Metacercariae-infected ants. *Journal of Parasitology, 102,* 495–500.

Hohorst, W., & Graefe, G. (1961). Ameisen-obligatorische Zwischenwirte des Lanzettegels (*Dicrocoelium dendriticum*). *Naturwissenschaften*, *48*, 229–30.

Krull, W. H., & Mapes, C. R. (1952). Studies on the biology of *Dicrocoelium dendriticum* (Rudolphi, 1819). Looss, 1899 (Trematoda: Dicrocoeliidae), including its relation to the intermediate host, *Cionella lubrica* (Muiller). VII. The second intermediate host of *Dicrocoelium dendriticum*. *Cornell Veterinarian*, *4*, 603–04.

Manga-González, M. Y. et al. (2001). Contributions to and review of dicrocoeliosis, with special reference to the intermediate hosts of *Dicrocoelium dendriticum*. *Parasitology*, *123*, 91–114.

Martin-Vega, D. et al. (2018). 3D virtual histology at the host/parasite interface: Visualisation of the master manipulator, *Dicrocoelium dendriticum*, in the brain of its ant host. *Scientific Reports*, *8*, 8587.

Tarry, D. W. (1969). *Dicrocoelium dendriticum*: The life cycle in Britain. *Journal of Helminthology*, *43*, 403–16.

ANTS PROTECT THEMSELVES AGAINST PATHOGENS

Clardy, J. et al. (2009). The natural history of antibiotics. *Current Biology*, *19*.

Cremer, S. et al. (2018). Social immunity: emergence and evolution of colony-level disease protection. *Annual Review of Entomology*, *63*, 105–23.

Schulz, T. R. (1999). Ants, plants and antibiotics. *Nature*, *398*, 747–48.

Stroeymeyt, N. et al. (2018). Social network plasticity decreases disease transmission in a eusocial insect. *Science*, *362*, 941–45.

SICK ANTS LEAVE THE NEST VOLUNTARILY

Bos, N. et al. (2011). Sick ants become unsociable. *Journal of Evolutionary Biology*, *25*, 342–51.

Chapuisat, M. (2010). Social evolution: Sick ants face death alone. *Current Biology*, *20*.

Heinze, J., & Walter, B. (2010). Moribund ants leave their nests to die in social isolation. *Current Biology*, *20*, 249–52.

ANTS LIVE IN HARMONY WITH ANTIBIOTIC-PRODUCING BACTERIA

Hedges, S. (1989). Science: Ant antibody fights fungal infections in humans. *New Scientist*, *1691*.

Ortega, H. E. et al. (2019). Antifungal compounds from Streptomyces associated with attine ants also inhibit Leishmania donovani. *PLoS Neglected Tropical Diseases*, *13*, e0007643.

Peakall, R. et al. (1990). The significance of ant and plant traits for ant pollination in *Leporella fimbriata*. *Oecologia*, *84*, 457–60.

Van Arnam, E. B. et al. (2016). Selvamicin, an atypical antifungal polyene from two alternative genomic contexts. *Proceedings of the National Academy of Sciences USA*, *113*, 12940–45.

MEGAPONERA ANALIS WORKERS RESCUE THEIR INJURED
SISTERS FROM THE BATTLEFIELD AFTER A RAID

Frank, E. T. et al. (2017). Saving the injured: Rescue behavior in
the termite-hunting ant *Megaponera analis*. *Science Advances, 3*,
e1602187.

Frank, E. T. et al. (2018). Wound treatment and selective help in a
termite-hunting ant. *Proceedings of the Royal Society B, 285.*

CHAPTER 12: THE PATH TO WORLD DOMINATION

THE TROPICAL FIRE ANT *SOLENOPSIS GEMINATA* WAS SPREAD ACROSS
THE WORLD IN THE SIXTEENTH CENTURY BY SPANISH SAILORS

Gotzek, D. et al. (2015). Global invasion history of the tropical fire ant:
A stowaway on the first global trade routes. *Molecular Ecology, 24*,
374–88.

McGlynn, T. P. (1999). The worldwide transfer of ants: geographical
distribution and ecological invasions. *Journal of Biogeography, 26*,
535–48.

THE ARGENTINE ANT *LINEPITHEMA HUMILE* REIGNS OVER THE
MEDITERRANEAN COAST OF WESTERN EUROPE WITH TWO
SUPERCOLONIES

Blight, O. et al. (2010). A new colony structure of the invasive Argentine
ant (*Linepithema humile*) in Southern Europe. *Biological Invasions,
12*, 1491–97.

Jaquiéry, J. et al. (2005). Multilevel genetic analyses of two European
supercolonies of the Argentine ant, *Linepithema humile*. *Molecular
Ecology, 14*, 589–98.

Roura-Pascual, N. et al. (2004). Geographical potential of Argentine
ants (*Linepithema humile* Mayr) in the face of global climate change.
Proceedings of the Royal Society B, 271, 1,557.

Suarez, A. V. et al. (1999). Behavioral and genetic differentiation
between native and introduced populations of the Argentine Ant.
Biological Invasions, 1, 43–53.

Tsutsui, N. D. et al. (2001). Relationships among native and introduced
populations of the Argentine ant (*Linepithema humile*) and the source
of introduced populations. *Molecular Ecology, 10*, 2151–61.

Wetterer, J. K. (2009). Worldwide spread of the Argentine ant,
Linepithema humile (Hymenoptera: Formicidae). *Myrmecological
News, 12*, 187–94.

THE GENETIC BOTTLENECK AND THE FOUNDER EFFECT
FACILITATE THE FORMATION OF SUPERCOLONIES

Jackson, D. E. (2007). Social evolution: Pathways to ant unicoloniality.
Current Biology, 17, R1063–R1064.

Lee, C. E. (2002). Evolutionary genetics of invasive species. *Trends in Ecology & Evolution, 17*, 386–91.

Moffett, M. W. (2012). Supercolonies of billions in an invasive ant: What is a society? *Behavioral Ecology, 23*, 925–33.

Suarez, A. V. et al. (2008). Genetics and behavior of a colonizing species: The invasive Argentine Ant. *The American Naturalist, 172*, 72–84.

THE YELLOW CRAZY ANT *ANOPLOLEPSIS GRACILIPES* IS DESTROYING THE ECOLOGICAL BALANCE ON CHRISTMAS ISLAND

Abbott, K. L. (2005). Supercolonies of the invasive yellow crazy ant, *Anoplolepis gracilipes*, on an oceanic island: Forager activity patterns, density and biomass. *Insectes Sociaux, 52*, 266–73.

Denslow, J. S. (2001). The ecology of insular biotas. *Trends in Ecology & Evolution, 16*, 423–24.

O'Dowd, D. J. et al. (2003). Invasional "meltdown" on an oceanic island. *Ecology Letters, 6*, 812–17.

THE RED IMPORTED FIRE ANT *SOLENOPSIS INVICTA* HAS CONQUERED THE SOUTHERN US

Ascunce, M. S. et al. (2011). Global invasion history of the fire ant *Solenopsis invicta*. *Science, 331*, 1066–68.

Morrison, L. W. (2002). Long-term impacts of an arthropod-community invasion by the imported fire ant, *Solenopsis invicta*. *Ecology, 83*, 2337–45.

Morrison, L. W. et al. (2004). Potential global range expansion of the invasive fire ant, *Solenopsis invicta*. *Biological Invasions, 6*, 183–91.

Tschinkel, W. R. (2006). *The Fire Ants*. Harvard University Press.

FORMICA FUSCOCINEREA HAS MIGRATED FROM AUSTRIA TO SOUTHERN GERMANY WITHOUT HUMAN INTERVENTION

Pohl, A. et al. (2018). Mass occurrence and dominant behavior of the european ant species *Formica fuscocinerea* (Forel). *Journal of Insect Behavior, 31*, 12–28.

THE INVASIVE GARDEN ANT *LASIUS NEGLECTUS* WAS FIRST IDENTIFIED AS A SPECIES IN 1990

Espadaler, X., & Rey, S. (2001). Biological constraints and colony founding in the polygynous invasive ant *Lasius neglectus* (Hymenoptera, Formicidae). *Insectes Sociaux, 48*, 159–64.

Seifert, B. (2000). Rapid range expansion in *Lasius neglectus* (Hymenoptera, Formicidae)—an Asian invader swamps Europe. *Deutsche Entomologische Zeitschrift, 47*, 173–79.

Tartally, A. et al. (2016). Collapse of the invasive garden ant, *Lasius neglectus*, populations in four European countries. *Biological Invasions, 18*, 3127–31.

THE TROPICAL PHARAOH ANT *MONOMORIUM PHARAONIS* SPREADS PATHOGENS IN CLINICS, KITCHENS, AND BAKERIES

Beatson, S. H. (1972). Pharaoh's ants as pathogen vectors in hospitals. *The Lancet, 299,* 425–27.

Bertelsmeier, C. et al. (2015). Worldwide ant invasions under climate change. *Biodiversity and Conservation, 24,* 117–28.

Gliniewicz, A. et al. (2006). Pest control and pesticide use in hospitals in Poland. *Orcis Indoor and Built Environment, 15,* 57–61.

Wetterer, J. K. (2010). Worldwide spread of the pharaoh ant, *Monomorium pharaonis* (Hymenoptera: Formicidae). *Myrmecological News, 13,* 115–29.

SOME INVASIVE SPECIES DISAPPEAR AGAIN OF THEIR OWN ACCORD

Cooling, M. et al. (2012). The widespread collapse of an invasive species: Argentine ants (*Linepithema humile*) in New Zealand. *Biology Letters, 8,* 430–33.

Cooling, M., & Hoffmann, B. D. (2015). Here today, gone tomorrow: declines and local extinctions of invasive ant populations in the absence of intervention. *Biological Invasions, 17,* 3351–57.

Lester, P. J., & Gruber, M. A. M. (2016). Booms, busts and population collapses in invasive ants. *Biological Invasions, 18,* 3091–3101.

Tartally, A. et al. (2016). Collapse of the invasive garden ant, *Lasius neglectus,* populations in four European countries. *Biological Invasions, 18,* 3127–31.

THE TAWNY CRAZY ANT *NYLANDERIA FULVA* IS DISPLACING THE RED IMPORTED FIRE ANT

Chen, J. et al. (2013). Defensive chemicals of tawny crazy ants, *Nylanderia fulva* (Hymenoptera: Formicidae) and their toxicity to red imported fire ants, *Solenopsis invicta* (Hymenoptera: Formicidae). *Toxicon, 76,* 160–66.

Kumar, S. et al. (2015). Evidence of niche shift and global invasion potential of the Tawny Crazy ant, *Nylanderia fulva. Ecology and Evolution, 5,* 4628–41.

LeBrun, E. G. et al. (2013). Imported crazy ant displaces imported fire ant, reduces and homogenizes grassland ant and arthropod assemblages. *Biological Invasions, 15,* 2429–42.

———. (2014). Chemical warfare among invaders: A detoxification interaction facilitates an ant invasion. *Science, 343,* 1014–17.

CHAPTER 13: CRAZY CRITTERS

AFRICAN *MELISSOTARSUS* ANTS ARE SO WELL-ADAPTED TO LIFE INSIDE TUNNELS, THEY CANNOT STAND ON FLAT GROUND

Buehler, J. (2018). Evolution turned this ant into a living drill. Sciencemag, sciencemag.org/news/2018/08/evolution-turned-ant-living-drill.

Khalife, A. et al. (2018). Skeletomuscular adaptations of head and legs of *Melissotarsus* ants for tunneling through living wood. *Frontiers in Zoology*, *15*, 1–11.

Mony, R. et al. (2013). *Melissotarsus* ants are likely able to digest plant polysaccharides. *Comptes Rendus Biologies*, *336*, 500–04.

CEPHALOTES ATRATUS AND OTHER SPECIES GLIDE BACK TO THE TREE FROM WHICH THEY HAVE FALLEN

Yanoviak, S. P. et al. (2005). Directed aerial descent in canopy ants. *Nature*, *433*, 624–26.

———. (2008). Directed aerial descent behavior in African canopy ants (Hymenoptera: Formicidae). *Journal of Insect Behavior*, *21*, 164–71.

———. (2010). Aerial maneuverability in wingless gliding ants (*Cephalotes atratus*). *Proceedings of the Royal Society B*, *277*, 1691.

TRAP-JAW ANTS USE THEIR MANDIBLES TO CATAPULT THEMSELVES INTO THE AIR

Gronenberg, W. et al. (1993). Fast trap jaws and giant-neurons in the ant *Odontomachus*. *Science*, *262*, 561–63.

Larabee, F. J., & Suarez, A. V. (2015). Mandible-powered escape jumps in trap-jaw ants increase survival rates during predator-prey encounters. *PLoS One*, *10*, e0124871.

Larabee, F. J. et al. (2017). Performance, morphology and control of power-amplified mandibles in the trap-jaw ant *Myrmoteras* (Hymenoptera: Formicidae). *Journal of Experimental Biology*, *220*, 3062–71.

Patek, S. N. et al. (2006). Multifunctionality and mechanical origins: Ballistic jaw propulsion in trap-jaw ants. *Proceedings of the National Academy of Sciences USA*, *103*, 12787–92.

THE MALES OF THE SOUTH AMERICAN SPECIES *HYPOPONERA OPACIOR* MATE WITH YOUNG QUEENS BEFORE THEY HAVE HATCHED FROM THEIR COCOONS

Foitzik, S. et al. (2002). Mate guarding and alternative reproductive tactics in the ant *Hypoponera opacior*. *Animal Behavior*, *63*, 597–604.

Kureck, I. M. et al. (2011). Wingless ant males adjust mate-guarding behaviour to the competitive situation in the nest. *Animal Behaviour*, *82*, 339–46.

MALE AND FEMALE LITTLE FIRE ANTS (*WASMANNIA AUROPUNCTATA*) ARE SO GENETICALLY DIFFERENT THAT THEY CAN APPEAR TO BE SEPARATE SPECIES

Foucaud, J. et al. (2009). Reproductive system, social organization, human disturbance and ecological dominance in native populations of the little fire ant, *Wasmannia auropunctata*. *Molecular Ecology*, *18*, 5059–73.

Fournier, D. et al. (2005). Clonal reproduction by males and females in the little fire ant. *Nature, 435,* 1230–34.

Queller, D. (2005). Evolutionary biology—Males from Mars. *Nature, 435,* 1167–68.

CORK-HEADED ANT GUARDS USE THEIR HEADS TO PLUG THE ENTRANCE TO THEIR NESTS

Fujioka, H. et al. (2019). Observation of plugging behaviour reveals entrance-guarding schedule of morphologically specialized caste in *Colobopsis nipponicus. Ethology, 125,* 526–34.

Hasegawa, E. (1993). Nest defense and early production of the major workers in the dimorphic ant *Colobopsis nipponicus* (Wheeler) (Hymenoptera, Formicidae). *Behavioral Ecology and Sociobiology, 33,* 73–77.

Laciny, A. et al. (2017). Morphological variation and mermithism in female castes of *Colobopsis* sp. nrSA, a Bornean "exploding ant" of the *Colobopsis cylindrica* group (Hymenoptera: Formicidae). *Myrmecological News, 24,* 91–106.

Walker, J., & Stamps, J. (1986). A test of optimal caste ration theory using the ant *Camponotus* (*Colobopsis*) *impressus. Ecology, 67,* 1052–62.

WORKERS OF THE BRAZILIAN SPECIES *FORELIUS PUSILLUS* SEAL THEIR NESTS FROM OUTSIDE IN THE EVENINGS AND RUN AWAY TO DIE

Bourke, A. F. G. (2008). Social evolution: Daily self-sacrifice by worker ants. *Current Biology, 18,* 1100–01.

Shorter, J. R., & Rueppell, O. (2012). A review on self-destructive defense behaviors in social insects. *Insectes Sociaux, 59,* 1–10.

Tofilski, A. T. et al. (2008). Preemptive defensive self-sacrifice by ant workers. *American Naturalist, 172,* 239–43.

CAMPONOTUS SAUNDERSI BLOWS ITSELF UP, SPREADING ITS POISON OVER ITS ATTACKER

Davidson, D. W. et al. (2012). Histology of structures used in territorial combat by Borneo's "exploding ants." *Acta Zoologica, 93*(4), 487–91.

Jones, T. H. et al. (2004). The chemistry of exploding ants, *Camponotus* SPP. (*Cylindricus* COMPLEX). *Journal of Chemical Ecology, 30*(8), 1479–92.

Laciny, A. et al. (2018). *Colobopsis explodens* sp. n., model species for studies on "exploding ants" (Hymenoptera, Formicidae), with biological notes and first illustrations of males of the *Colobopsis cylindrica* group. *ZooKeys, 751,* 1–40.

COMPUTER SCIENTISTS MIMIC THE PHEROMONE TRAILS OF ANTS TO FIND THE BEST ONLINE PATHWAYS

Bonabeau, E. et al. (2000). Inspiration for optimization from social insect behaviour. *Nature, 406,* 39–42.

Musco, C. et al. (2017). Ant-inspired density estimation via random walks. *Proceedings of the National Academy of Sciences USA, 114,* 10534–41.

Werfel, J. et al. (2014). Designing collective behavior in a termite-inspired robot construction team. *Science, 343,* 754–58.

IN HEAVY RAIN, *TETRAPONERA BINGHAMI* AND *CATAULACUS MUTICUS* DRINK THE WATER THAT FLOODS INTO THEIR NESTS AND SPIT OR PEE IT OUT OUTSIDE THE NEST ENTRANCE

Kolay, S., & Annagiri, S. (2015). Dual response to nest flooding during monsoon in an Indian ant. *Scientific Reports, 5,* 13716.

LeBrun, E. G. et al. (2011). Convergent evolution of levee building behavior among distantly related ant species in a floodplain ant assemblage. *Insectes Sociaux, 58,* 263–69.

Maschwitz, U., & Moog, J. (2000). Communal peeing: a new mode of flood control in ants. *Naturwissenschaften, 87,* 563–65.

PHOTO CREDITS

INDEX

Page numbers in *italics* refer to photographs, watercolors, and their captions.

ABOUT THE AUTHORS

SUSANNE FOITZIK is an evolutionary biologist, behavioral scientist, and international authority on ants. After completing her PhD in ant evolution and behavior and conducting postdoctoral work in the US, she became a professor at Ludwig Maximilian University of Munich. She now holds a chair at Johannes Gutenberg University in Mainz, Germany, where she studies the behaviors and social evolution of ants, with a particular focus on slavemaking ants. Her findings have been published in more than 110 scientific papers.

OLAF FRITSCHE is a science journalist and biophysicist with a PhD in biology. He was previously an editor at the German-language edition of *Scientific American,* is the author and coauthor of many books, and has been published in a wide variety of newspapers and magazines. He lives in Germany.